DATE DUE

GAYLORD			PRINTED IN U.S.A.

U. Seiffert · L.C. Jain · P. Schweizer (Eds.)

Bioinformatics Using Computational Intelligence Paradigms

Studies in Fuzziness and Soft Computing, Volume 176

Editor-in-chief
Prof. Janusz Kacprzyk
Systems Research Institute
Polish Academy of Sciences
ul. Newelska 6
01-447 Warsaw, Poland
E-mail: kacprzyk@ibspan.waw.pl

Further volumes of this series can be found on our homepage: springeronline.com

Vol. 159. E. Damiani, L.C. Jain, M. Madravia (Eds.)
Soft Computing in Software Engineering, 2004
ISBN 3-540-22030-5

Vol. 160. K.K. Dompere
Cost-Benefit Analysis and the Theory of Fuzzy Decisions – Fuzzy Value Theory, 2004
ISBN 3-540-22161-1

Vol. 161. N. Nedjah, L. de Macedo Mourelle (Eds.)
Evolvable Machines, 2005
ISBN 3-540-22905-1

Vol. 162. R. Khosla, N. Ichalkaranje, L.C. Jain
Design of Intelligent Multi-Agent Systems, 2005
ISBN 3-540-22913-2

Vol. 163. A. Ghosh, L.C. Jain (Eds.)
Evolutionary Computation in Data Mining, 2005
ISBN 3-540-22370-3

Vol. 164. M. Nikravesh, L.A. Zadeh, J. Kacprzyk (Eds.)
Soft Computing for Information Prodessing and Analysis, 2005
ISBN 3-540-22930-2

Vol. 165. A.F. Rocha, E. Massad, A. Pereira Jr.
The Brain: From Fuzzy Arithmetic to Quantum Computing, 2005
ISBN 3-540-21858-0

Vol. 166. W.E. Hart, N. Krasnogor, J.E. Smith (Eds.)
Recent Advances in Memetic Algorithms, 2005
ISBN 3-540-22904-3

Vol. 167. Y. Jin (Ed.)
Knowledge Incorporation in Evolutionary Computation, 2005
ISBN 3-540-22902-7

Vol. 168. Y. P. Tan, K. H. Yap, L. Wang (Eds.)
Intelligent Multimedia Processing with Soft Computing, 2004
ISBN 3-540-23053-X

Vol. 169. C.R. Bector, S. Chandra
Fuzzy Mathematical Programming and Fuzzy Matrix Games, 2004
ISBN 3-540-23729-1

Vol. 170. M. Pelikan
Hierarchical Bayesian Optimization Algorithm, 2005
ISBN 3-540-23774-7

Vol. 171. J.J. Buckley
Simulating Fuzzy Systems, 2005
ISBN 3-540-24116-7

Vol. 172. P. Melin, O. Castillo
Hybrid Intelligent Systems for Pattern Recognition Using Soft Computing, 2005
ISBN 3-540-24121-3

Vol. 173. B. Gabrys, K. Leiviskä, J. Strackeljan (Eds.)
Do Smart Adaptive Systems Exist?, 2005
ISBN 3-540-24077-2

Vol. 174. M. Negoita, D. Neagu, V. Palade
Computational Intelligence, 2005
ISBN 3-540-23219-2

Vol. 175. A.M. Gil-Lafuente
Fuzzy Logic in Financial Analysis, 2005
ISBN 3-540-23213-3

U. Seiffert
L.C. Jain
P. Schweizer
(Eds.)

Bioinformatics Using Computational Intelligence Paradigms

With 69 Figures and 20 Tables

Dr. Udo Seiffert
Leibniz-Institute of Plant Genetics and Crop Plant Research
Pattern Recognition
Correnstr. 3
06466 Gatersleben
Germany

Professor Dr. Lakhmi C. Jain
University of South Australia
Knowledge-Based Intelligent Engineering Systems Centre
Mawson Lakes
5095 Adelaide
SA, Australia

Dr. Patrick Schweizer
Leibniz-Institute of Plant Genetics and Crop Plant Research
Transcriptome Analysis
Correnstr. 3
06466 Gatersleben
Germany

ISSN 1434-9922
ISBN 3-540-22901-9 Springer Berlin Heidelberg New York

Library of Congress Control Number: 2004117076

Springer is a part of Springer Science+Business Media
springeronline.com

Printed in Germany

Typesetting: Data conversion by the author.
Final processing by PTP-Berlin Protago-TEX-Production GmbH, Germany
Cover-Design: Erich Kirchner, Heidelberg
Printed on acid-free paper 89/3020/Yu - 5 4 3 2 1 0

Preface

Bioinformatics is undoubtedly a remarkably fast growing field of research and real-world applications. It is considered to have enormous potential for current and future generations and the development of all of mankind. However, after an initial phase with enthusiasm similar to that during the gold-rush, people have now come to a point where they realize that many problems are still far away from being solved. This obviously comes from the high complexity of these problems, the relatively extensive and expensive equipment which is necessary to perform in-depth investigations, and last but not least, from the interdisciplinary character of this special field.

On the other hand, this interdisciplinary environment of life science, computer science, mathematics, engineering and other disciplines makes the field fascinating and provides a platform for working on the edge of a real breakthrough. As the two components of the name suggest, Bioinformatics deals with the use of computers in biology-related sciences. However, beyond this, the application of biologically inspired information processing on a computer, as covered by the term Computational Intelligence, joins together these two parts in a different but by no means less interesting manner. Computational Intelligence has evolved to a widely utilized and accepted bunch of tools to accompany, improve or even substitute conventional algorithms and information processing procedures. The book combines Bioinformatics and Computational Intelligence, leading to an interdisciplinary cross-fertilization.

The editors would like to take this opportunity to express their gratitude to all the authors, without whose hard work this book would not have been possible. Furthermore we would like to thank the staff of Springer Verlag for their support in preparing this book.

Gatersleben, Germany and Adelaide, Australia,
July 2004

Udo Seiffert
Lakhmi C. Jain
Patrick Schweizer

Contents

Medical Bioinformatics: Detecting Molecular Diseases with Case-Based Reasoning

Ralf Hofestädt and Thoralf Töpel

Bioinformatics Department, Faculty of Technology, Bielefeld University, Germany
ralf.hofestaedt@uni-bielefeld.de

Summary. Based on the Human Genome Project, the new interdisciplinary subject of bioinformatics has become an important research topic during the last decade. A catalytic element of this process is that the methods of molecular biology (DNA sequencing, proteomics etc.) allow the automated generation of data from cellular components. Based on this technology, robots are able to sequence small genomes in a few weeks. Moreover, the semi-automatic assembly and annotation of the sequence data can only be done using the methods of computer science. To handle massive amount of data using hardware and software is one reason for the actual success of bioinformatics (Collado-Vides and Hofestädt 2002). Today, besides the genome, protein and pathway data, a new domain of data is arising – the so-called proteomic project, which allows the identification of specific protein profiles in concentration and location.

1 Bioinformatics – Definition and Overview

Molecular data is stored in database systems available via the Internet. Based on that data, different questions can be answered using specific analysis tools. For DNA sequences, we are looking for powerful software tools which will be able to predict DNA-functional units. Today this topic is called "From the Sequence to the Function" or "Post Genomics". The most important application area of this new research topic is molecular medicine. Therefore bioinformatics has also become an important topic for medical informatics and medicine. Regarding definitions of bioinformatics in the WWW or literature we can see different views: The German definition is a global definition (Hofestädt et al. 1993). On one hand, the application of the methods and concepts of computer science in biology represents the main focus. On the other hand, looking into the history of computer science, we can identify important innovations coming directly from the analysis of molecular information processing systems. Regarding this aspect, we can identify important innovations arising from the analysis of molecular mechanisms. Thus aspect, we can distinguish between direct and indirect innovations. The implementation of genetic algorithms, neural networks or DNA computing mechanisms try to solve hard problems using molecular heuristics. This topic can be called biological paradigm of computation. Moreover,

the definition of formal systems, like the finite automaton (H. Kleene), the cellular automaton (J. v. Neumann) or L-System (A. Lindenmayer), is based on the idea of the implementation of analysis tools for modelling neural networks.
The common definition of bioinformatics addresses the application of methods and concepts of computer science in the field of biology. Bioinformatics currently addresses three main topics. The first major topic is sequence analysis or genome informatics. Its basic tasks are: assembling sequence fragments, automatic annotation, pattern matching and implementation of database systems, like EMBL, TRANSFAC, PIR, RAMEDIS, KEGG etc.. The sequence alignment problem still represents the kernel of sequence analysis tools. Nevertheless, sequence analysis is not a new topic. It was, and still is, a topic of theoretical biology and computational biology. Protein design is the second current major research topic of bioinformatics. The first task was to implement information systems that represent knowledge about the proteins. Today many different systems, like PIR or PDB, are available. The main goal of this research topic is to develop useful models which will allow the automatic calculation of 3D structures, including the prediction of the molecular behavior of this protein. Until now, molecular modelling failed. Protein design is also not a new research topic. Its roots are coming from biophysics, pharmaco kinetics and theoretical biology. The third major topic is metabolic engineering. Its goal is the analysis and synthesis of metabolic processes. The basic molecular information of metabolic pathways is stored in database systems, like KEGG, WIT, etc.

1.1 Virtual Cell

The idea of metabolic engineering follows from the idea of the virtual cell (Collado-Vides and Hofestädt 2002). Using molecular data and knowledge, the definition of specific models allows the implementation of simulation tools for cellular processes. Behind the algorithmic analysis of molecular data, modelling and simulation methods and concepts allow the analysis and synthesis of complex gene controlled metabolic networks (Kolchanov and Hofestädt 2003). The actual data and knowledge of the structure and function of molecular systems is still rudimentary. Furthermore, the experimental data available in molecular databases has a high error rate, while biological knowledge has a high rate of uncertainty. Therefore, modelling and simulation and methods of artificial intelligence are useful in addressing important questions. Such a formal description can be used to specify a simulation environment. Modelling and simulation can thus be interpreted as the basic step for implementing virtual worlds that allow virtual experiments.

As already mentioned, more than 500 database and information systems are available, representing molecular knowledge. Furthermore, a lot of analysis tools and simulation environments are available. That means that basic components of the electronic infrastructure for the implementation of a virtual cell is present. The concepts and tools which are available in the literature and the internet are based on specific questions, such as the gene regulation process phenomena, or the biochemical process control. To solve current questions, we have to implement integrative tools (integrative bioinformatics) which can be used to implement a virtual cell. One of the first implementations is the E-Cell project of M. Tomita (http://www.e-cell.org). Many new virtual cell projects are following the E-Cell project. However, as already mentioned it will take a long period of time to implement a useful and powerful virtual cell. Rudimentary knowledge is one problem confronting the implementation of such system. Moreover, molecular data and knowledge about the logic of molecular processes are still missing. Until now we are still not able to understand the quantitative behavior of simple metabolic processes.

1.2 Benefits

Bioinformatic activities will implement the electronic infrastructure of molecular biology, helping to analyze molecular and biochemical mechanisms and systems. Moreover, these methods will support not only the analysis process in this important branch of science. Supporting biotechnology means that the application spectrum of bioinformatics is diverse. At present supporting drug design, molecular diagnosis and gene therapy are more or less the most important applications. The detection of metabolic diseases can be done using information systems in combination with methods of artificial intelligence. Therefore, information and expert systems for the detection of metabolic diseases as well as tools for the analysis of genotype/phenotype correlations have already been implemented. In our paper we will focus to this topic of medical bioinformatics. Thus, the future of molecular medicine will depend on the future of molecular biology.

1.3 Barriers

Bioinformatics methods in use are database systems, information systems, visualization, animation, modelling, and simulation. Evaluation studies of molecular database systems show that most systems contain a high proportion of incorrect and/or junk data. Moreover, the copy and transfer process from the selected and relevant papers to the database, which is done by humans, also shows a high error rate, and it will not be easy to solve this problem in the near future. Another problem is that we are not able to implement analysis tools for a lot of important questions. We need clear and exact definitions and specifications for the development of useful analysis tools. This is often not the case in the field of biology (medicine too). For example the fundamental term "homology of sequences" has different definitions. This is one reason for so many different alignment algorithms. The other reason is the high complexity of time and space for many of these problems. Complexity is the main argument against the implementation of the virtual cell within the next few decades. Finally, the main barriers come directly from molecular biology. Today, it seems as though we will never understand basic molecular mechanisms, such as the fundamental process of gene regulation.

1.4 Prospects for the Near Future

Different and powerful information systems for scientists, patients and doctors to represent the basic knowledge of molecular biology are already in use. Moreover, the relevant data is growing exponential and distributed over local networks or over the whole world (internet). The medical, clinical and molecular data can be used to support therapy, diagnosis and drug design. The molecular diagnosis of metabolic diseases is a current research topic. Thousands of metabolic diseases are known and about 500 relevant inherited errors are discussed in the literature.

Based on medical data of inherited errors, the German Human Genome Project initiated a project, to discuss the actual benefits of molecular information fusion in combination with modeling and simulation methods (Mischke et al. 2001, Döhr et al. 2002). Databases such as METAGENE, KEGG, BRENDA, TRANSFAC, TRANSPATH and RAMEDIS will be integrated into this project. RAMEDIS was developed and implemented as one part of this project. Regarding RAMEDIS it can be shown that also the integration of AI methods is important for further projects in Bioinformatics.

2 An Introduction to Case-Based Reasoning

2.1 Motivation

Expert or knowledge-based systems are already part of successful artificial intelligence research. Most of these systems use rules, frames or clauses to formalize stored information. But an intricate process of knowledge elicitation is required that is often referred to as the knowledge elicitation bottleneck. Other problems in developing knowledge-based decision support systems are the required special skills and the difficult maintenance of the implemented system (Watson 1995).

In fact, the specialists require expert knowledge by contact with similar problem over many years. A professional keeps in mind the resulting conclusions, applied methods and practices particularly in context of solved tasks. According to this, the case-based reasoning (CBR) approach uses old experiences to understand and solve new problems (Kolodner 1992). The CBR system remembers a previous situation similar to the current one and can adapting a previously successful solution to that problem. This procedure of retrieving and adapting existing problem solving policies has several advantages (Watson 1995).

- The CBR system does not require an explicit domain model and the so far intricate elicitation process becomes a simple task of collecting case histories. Existing sets of data about cases, e.g. electronically stored in patient records, can be directly incorporated to the CBR system.
- The implementation process of the CBR system is mainly reduced to an identification of significant features that describe a case.
- The application of database techniques enables the CBR system to manage large volumes of information.
- The maintenance of the CBR system is easier because new knowledge is acquired as cases.

One point of origin of the CBR approach is the exploration of the computer-based support in the use of episodic expert knowledge in (Schank 1982). Basing on results from cognition psychology, a theory of understanding, remembering and learning of experiences was established. The notion *Dynamic Memory* described the dependence of this method on the human memory. Starting from that basic principle a lot of theoretic and application research resulted. The following sections present techniques and benefits of the CBR approach.

2.2 Case-Based Reasoning Techniques

As described before the CBR approach is based on stored cases that represent previously experienced problem situations. New problems are solved by searching similar situations in the case base and adapting their solution to the new problem. This process of case-based reasoning can be described as a cycle with four single steps (Aamodt and Plaza 1994) and is illustrated in Fig. 1.

Before explaining the steps of the process cycle several terms or definitions should be outlined. A *case* usually characterizes a problem situation. This case consists of a set of 2-tuples with a feature and a corresponding feature characteristic. A problem situation comprehends only features represented in this set. The data type of the feature characteristic depends on the type of the feature describing the condition of a examined system, e.g. yes/no, numeric or multiple choice. In many cases each feature can take independently from the current data type the value 'unknown'. But not all features must be explicitly part of a case, in a step of data preparation so called feature interpretations can be extracted from other retrieved features.

Consequently, a known problem situation and its corresponding solution is stored in the *case base* also known as *knowledge base* of the CBR system. A new case or unsolved case describes a new problem situation. This terms are the starting point of the cyclic process of CBR comprising four basis principles.

1. Retrieve
   A new problem situation is described as a *new case* and imported to the CBR system. In interaction with the case base as a collection of previous cases the CBR system searches for one or more similar cases (*retrieved case*) by using appropriate similarity measures.
2. Reuse
   By the combination of the *retrieved case* and the *new case* a *solved case* results that envelopes a proposed solution to the current problem statement.
3. Revise
   The proposed solution within the *solved case* is evaluated of its coverage to the problem statement and corrected if required. Result of the revise process is a *tested case*.
4. Retain
   To supply helpful experiences for future use, a *tested case* can be incorporated to the case base as a *learned case* or by modifying existing cases.

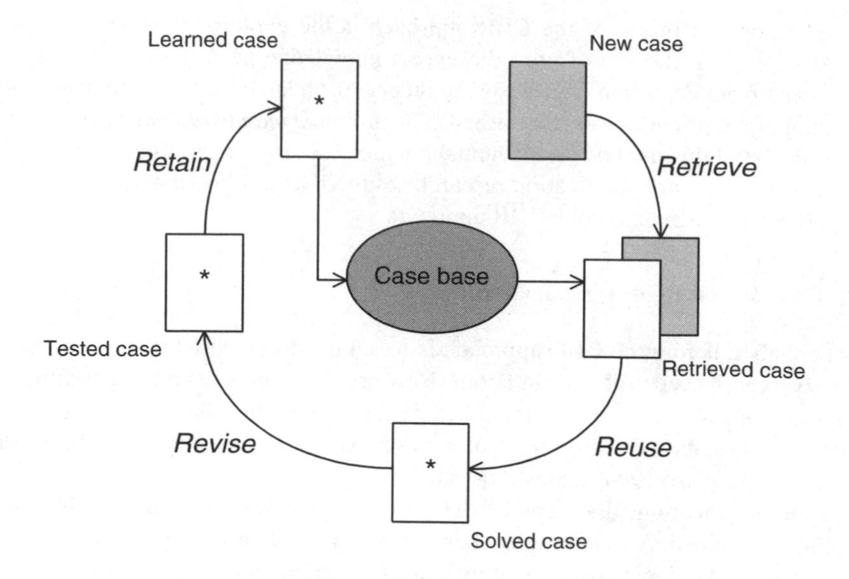

Fig. 1. The four work steps of the CBR cycle (adapted from Aamodt and Plaza, 1994).

In practice not all steps of the CBR cycle are applied to problem situations. Many applications support the user automatically in case retrieval and reuse. This graded implementation of the CBR principle can be basically diveded into two approaches.

- Case-match CBR
  This approach searches in the case base for the most similar case to the current problem situation and decides about the transfer of the solution to the problem situation.
- Case-adaptation CBR
  In addition to the case-match CBR the system adjusts the partly fitting solution of the most similar case to the current problem definition by additional domain knowledge.

In addition, the processing concept of the CBR cycle with four steps was refined in different publications. Thereby, single steps can overlap with the process model.

1. Preselection of a rough selection of suitable cases (potential solution candidates) from the case base as basis for the further steps
2. Evaluation of the similarity of the preselected cases
3. Adaption of the solutions of the most similar cases
4. Test and criticisation of the found solution
5. Evaluation of the results
6. Learning

The various concepts of the CBR process model are supported by special algorithms and methods. The support of the retieve process by a suitable indexation of the stored cases is only one example of this assisting methods. However, no further remarks to special algorithms and methods should be made here. More about these topics can be found in the appropriate literature.

2.3 Concluding Remarks

Within the research area of Artificial Intelligence the case-based reasoning approach offers a number of advantages for decision support in knowledge acquisition and system implementation. This approach bases on remembering a previous situation as a retrieved case similar to the current problem statement and adapting a previously successful solution to that problem. Particularly important is the ability of CBR systems to act as knowledge repositories of conclusions, applied methods and practices that easily can grow by learning new experiences. The occurrence of solutions and the behaviour of the system is comfortable to observe by recounting the similarity measures of retrieved cases.

Many of different problems can be simply solved by applying the process model of the CBR cycle. New problems are solved by retrieving previously experienced cases, reusing the case in content of the current problem situation, revising its proposed solution and retaining the new experience by incorporating it into the case base (Aamodt and Plaza 1994).

3 A Database for Mutations and Associated Phenotypes

3.1 Background

A lot of medical publications, for example given in proceedings, often present case reports of rare metabolic diseases. They represent valuable information for better understanding of rare conditions. Such inherited errors of metabolism are characterised by a block in a metabolic pathway, a deficiency of a transport protein or a defect in a storage mechanism caused by a gene defect. The originating defective gene leads up to an absent or wrong production of essential proteins, especially enzymes. But these enzymes that are important components of the biochemical processes in cells and tissues. They enable, disable or catalyse or regulate the biochemical reactions of metabolic pathways. Thus, these disorders of the metabolism result in a threatening deficiency or accumulation of metabolic intermediates in the human organism and their corresponding symptoms. If a patient is suspected of having an inherited error of metabolism, specialised biochemical laboratories analyse enzyme activities in specimen of different tissues (skin, liver etc.) and investigate body liquids as blood, urine etc. for unusual metabolic pattern. With molecular methods it is also possible to confirm a diagnosis and to define the defect gene. In a screening procedure a number of inherited metabolic errors can be already examined prenatally or immediately after birth, e.g. Phenylketonuria (PKU).

However, most of the information about a patient, his/her symptoms and treatments, may be lost by using common ways of publication. This section outlines the considerations for the design and development of a general approach for computer-based collection, storage and analysis of data about patients with inherited errors of metabolism. During these investigations a number of recommendations have been formulated that were converted to an appropriate architecture and implemented in a web-based prototype.

3.2 Requirements and Architecture

The related works like PAHdb (Scriver et al 2003), ARPKD and BioDef (Blau 1996) are widely accepted solutions to collect and store data of mutations and associated phenotypes in their special subject, but they are not suitable for use in a general approach. Other more theoretical works (Scriver et al. 2000, Porter et al. 2000) analyse guidelines and recommendations of mutation databases regarding the users – the human genomic community. As a result of the examination of already existing attempts, the following list of requirements has been identified.

- Comprehensive case reports
  The data of the different patients managed by the system has to be organised in the form of comprehensive case reports. Required parameters for the description of the single case, its therapy and aethiopathology have to be made available by the user.
- Data collection by health professionals
  Input and maintenance of the data of mutations and accompanying parameters takes place by the health professional. For that purpose an appropriate graphical user interface and access to advanced services is certainly helping to support and motivate users for inputs.
- Data input and data analysis
  To support users, a number of adequate software tools has to be made available that allow input and analysis of the information.
- Data privacy and data protection
  When working with electronically processible patient data it is important to guarantee the protection of data privacy. This requirement has to be kept in mind during the design procedure of the system architecture. At run time of the collection process attendant measures like patients declaration of consent and anonymisation protect data privacy too.
  Write access on stored case reports is reserved for the original author of the report. Additionally, the usual requirements for data privacy and data protection in reference to the keeping of data has to be complied by the database management system.
- Disease independent applicability
  The system has to be designed to cover as many diseases as possible caused by gene defects.

- Extensibility
  To extend the database to specific characteristics of case report, e.g. new lab findings, which were not considered during the initialization of the software, a suitable application has to be provided. The maintenance of this additional parameters is only available for users with advanced authorizations.
- High comparability of stored data
  During the process of data input of case reports by the user, the submission of synonymous and homonymous terms of medical definitions has to be prevented. This results in a high comparability of the stored data.
- Platform independence, world wide access, intuitive usability
  To activate a preferably high number of health professionals to insert and maintain data in the form of case reports, the system has to be platform independent and usable world wide via the internet. A intuitive usable graphical user interface simplifies and supports the data input and maintenance procedures for the health professionals.
  On the basis of the evaluation of related works and the compilation of the requirements of a general approach for collecting and analyzing mutations and associated phenotypes of diseases caused by gene defects, an architecture of an information system prototype has been developed. The following figure shows the architecture in form of an UML component diagram with four main partitions.
- Users
  Different users simultaneously access the system to utilize the different services provided by the just introduced architecture. With various levels of user rights the system organisms the permission or rejection of single user demands to the services.
- Data input component
  A software component for data input realists with three different modules the functions to maintain the data stock of the system. The administrator creates or deletes the data for the access of the authors to the services with the *Author Management*. Afterwards, the authors use the *Data Input Tool* to insert new mutations and associated phenotypes in form of case reports to the database. If necessary, new parameters for the description of a case report can be added to the database by using the *Parameter Management*.
- Data analysis component
  To analyst the data of the mutations and associated phenotypes stored as case reports in the database, a *Data Analysis Tool* is provided to support the user exploring and reading the data stock. By using the *Search Module*, simple queries on single parameters of the case reports or combined questions with multiple parameters in terms of case-based reasoning are possible. A successful interaction with the user results in a *Case Report Preparation* that produces the single case report within the graphical environment. This presentation is completed by the visualization of growth parameters (length, weight and body-mass index) and lab findings. Interested parties register with the *Author Management* to apply for a permit to insert and maintain mutation and phenotype data with the information

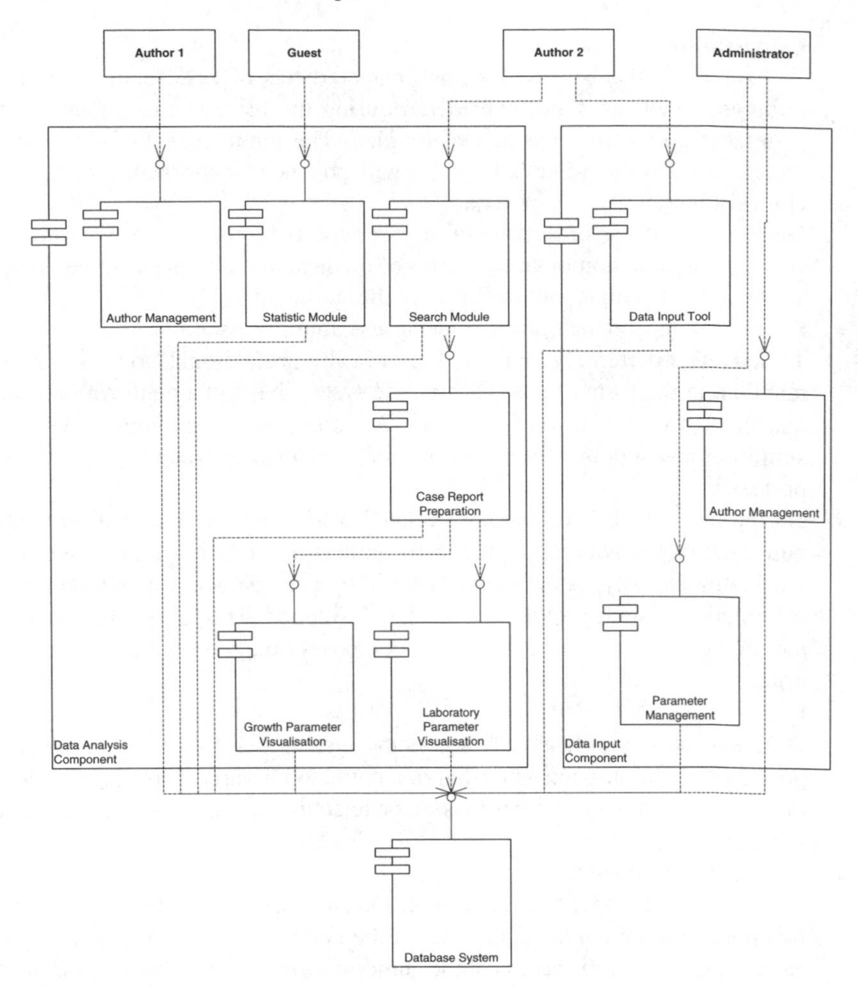

Fig. 2. Architecture of the RAMEDIS system as UML component diagram.

system. A simple Statistic Module presents general information on the current data stock, e.g. the number of available case reports and the according parameters.

- Database management system
  A professional database management system that complies with the fundamental requirements on non-redundant data management, transaction management and data security etc. is used to store the appearing data and make it available on applications demands.

The conceptual design of the database depends on the defined requirements before.

3.3 Database and Applications

The RAMEDIS system (Mischke et al. 2001) was implemented as a platform independent, web-based information system for rare metabolic diseases on the basis of separate case descriptions. From its launch three years ago about 600 anonymous patients have been published by physicians with 80 different genetic metabolic diseases with extensive details, e.g. about occurring symptoms and laboratory findings. Furthermore, since molecular data are also included as far as available, this information system will be helpful for genotype-phenotype investigations. Table 1 shows a summary of the current data stock of RAMEDIS.

By using largely standardized medical terms and conditions the content of the database is easy to compare and to analyst. In contrast to commonly used proceedings or electronic journal papers and databases on rare metabolic diseases our approach collects the data of different genetic metabolic diseases in a collective data model. In addition, a convenient graphical user interface is provided to the user for accessing the underlying information.

Table 1. Summary of data stock of RAMEDIS (as of February 2004)

Authors	51
Case reports	661
Number of predefined parameters ordered by areas	
Diagnoses	364
Symptoms	663
Lab findings	1350
Therapies and developments	83
Diets and drugs	144
Number of transmitted parameters ordered by areas	
Symptoms	3527
Symptoms per case	5.34
Lab findings	16347
Lab findings per case	24.73
Drugs and developments	2979
Diets and drugs	1482
Pictures and images	71
Molecular genetics	263

As already described with the conception of the architectural components RAMEDIS can be used on two different ways. The Data Input Tool enables the user to insert new patients and its according data to the database; thereby the right to edit information remains with the author that is identified by user login. An easily useable Java application was developed to support the user during the data input. Suitable data structures (e.g. trees) are used that guarantee a fast finding of necessary examination parameters within hundreds of variable traits after a graphic conversion. The essential verification of the data on consistency and plausibility are accomplished here in order to ensure a high data quality.

Most of clinical studies need only a small, well defined set of characteristics, which have to be committed rapidly. So an additional software module provides study specific input masks embedded in an appropriate security architecture usable with a common web browser. After the required log in with user name and corresponding password an user specific selection of input forms is presented. Thereby, one disease can offers several different forms depending on the type of the examination, e.g. basic data, history, clinical status or treatment. The data will be recorded in the database with its according author and examination date. This simplified approach of reduced selection enables higher performance of the data input procedure, which is especially important during the clinical practice.

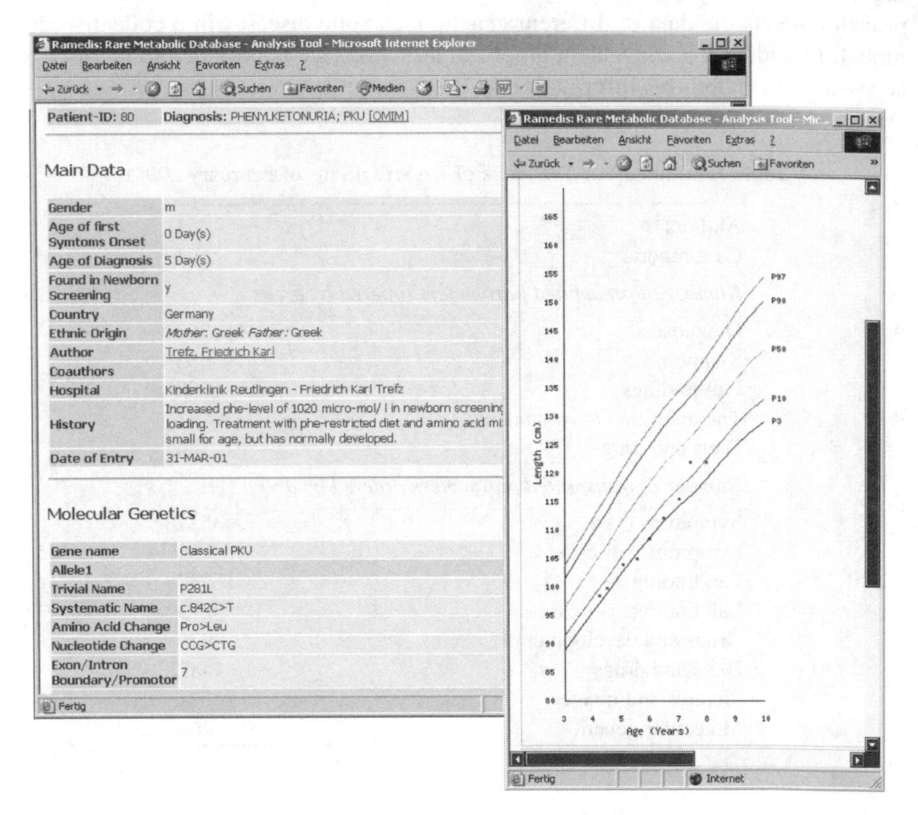

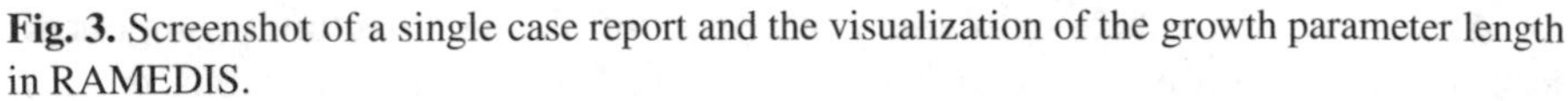

Fig. 3. Screenshot of a single case report and the visualization of the growth parameter length in RAMEDIS.

To browse, analyze and visualize the stored data, the Data Analysis Tool is also used with a common web browser like Netscape Navigator or MS Internet Explorer. Over a graphical user interface an extensive search form is offered that makes it possible to narrow the available case reports down to interesting attributes. After the selection of a certain case the collected data of symptoms, laboratory tests, treatment, etc. are presented. A screenshot of this software module shows Fig. 3. The large window on the left side presents main data and molecular genetics of a single case reports. On the right hand the growth parameter length is visualized within statistical data from a normal German population (Kromeyer-Hauschild et al. 2001). By dint of this, the growth of an ill patient in childhood can be compared to reference data extracted from several published examinations. This software applications increase the value of the collected data, simplifies their evaluation and motivates health professionals to input further case reports to the system.

Besides these software tools, the internet domain www.ramedis.de provides more information about the project, e.g. an extensive tutorial for handling the supplied tools.

3.4 Concluding Remarks

In this section we presented an approach to collect case reports of patients with rare metabolic defects world wide via the internet. An extensive and platform independent information system with the name RAMEDIS was developed that supports this task for an extendible number of different genetic diseases and enables co-operative studies. On the basis of a recommendations list an architecture for such a system was developed and prototypically implemented. A number of free-of-charge available software application support health professionals and other interested users in collecting, maintaining and analysing case reports with platform independent tools equipped with graphical user interfaces.

4 Assembling the Parts: CBR on Medical Data

4.1 Introduction

In the preceding chapters both the CBR approach for retrieving previously experienced cases and reusing these cases in content of the current problem situation and RAMEDIS as a database for mutations and associated phenotypes were presented. The following sections show how these two approaches are assembled to support electronically the detection of molecular diseases on real but anonymous patient data.

4.2 Method and Application

Based on the process model of the CBR cycle, illustrated in Fig. 1, an algorithm to retrieve similar cases from the collected patient data in RAMDIS and appropriate similarity measures to compare various cases were designed.

As a simple example of using the CBR approach with existing data, a web-based form was created within the graphical user interface that presents the parameters (features) symptom, laboratory finding and ethnical origin for querying in the volume of data. Fig. 4 shows a screenshot of this interface with the form fields specifying the feature characteristics of the current problem situation. By using a special selection window – shown in the figure on the right side – symptoms and laboratory findings can be selected within the volume of data. Additionally, the weight of the single features for the computation of the overall similarity can be defined.

After describing the current problem situation with the web-based form, in an intermediate step the appropriate, potentially similar data within the database (case base) are examined and put down in a temporary table with its corresponding similarity. This algorithm is described by the following steps.

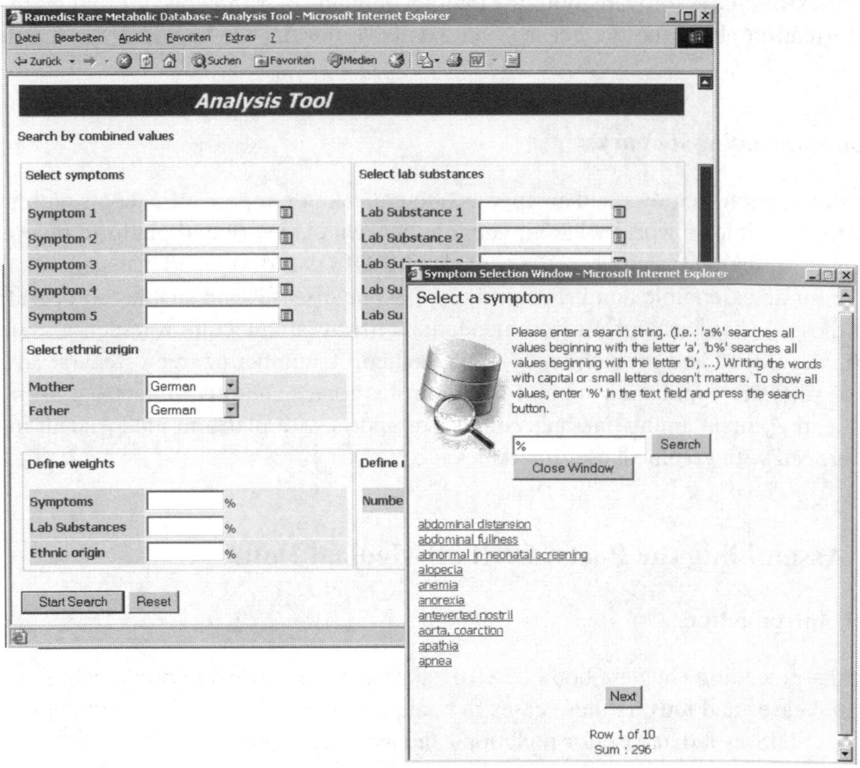

Fig. 4. Screenshot of the graphical user interface for defining feature weights, the features symptom, lab finding substance and ethnic origin and its representation of a similar problem situation.

1. Definition of a unique session identification
2. Computation of the summation for the maximum weights
3. Preselection of the similar cases from the database (case base)
4. For each preselected case

- Computation of the weights of the symptoms
- Computation of the weights of the laboratory findings
- Computation of the weights of the ethnical origin
- Computation of the overall case similarity
- Temporary storage of the case information with session id and similarity

5. Output of the computed similarities for each case with matching features

The result of this processing step is a list of similar cases with its respective similarity in comparison to the previously specified problem situation as shown in Fig. 5. In addition to the corresponding similarity of the itemized cases the matching features are specified. In the available example a query was constructed for the symptom 'Fever' in combination with the measurement of the laboratory substance 'Phenylalanine'. The first column contains the identification of the linked case in the database to receive more information about the certain patient with additional examination results and treatment directions. A preview of this case specific data gives the second column with diagnosis name of the retrieved case. The following three columns cover the matching features in comparison to the search case. The similarity of the case is displayed in the last column of the table.

4.3 Concluding Remarks

In this chapter an example for combining computational intelligence in the form of case-based reasoning with existing patient data was presented. Therefore graphical user interface was implemented to support the user. A robust algorithm gathers potentially similar case reports from the database and computes the respective similarity. The results of this process are presented to the user as a list with additional information for understanding the computation of the similarity and links to the corresponding case reports.

5 Conclusion and Future Work

We have presented our approach using principles of case-based reasoning for detecting molecular diseases. An introduction chapter outlined the basics for the use of case-based reasoning (CBR) with a number of continuative literature references. As a further part of this work, a prototype of an information system for collecting case reports of patients with rare metabolic defects world wide via the internet was established. This extensive and platform independent information system with the name RAMEDIS was developed to support this task for an extendible number of different genetic diseases and to enable co-operative studies. The database of case reports of this information system was used here to show the combination of CBR principles with existing data. Future work may include the analysis of further similarity measures and their fitting within the application.

Inherited errors of metabolism are caused by genetic defects that will cause a block in a metabolic pathway, a deficiency of a transport protein or a defect in a

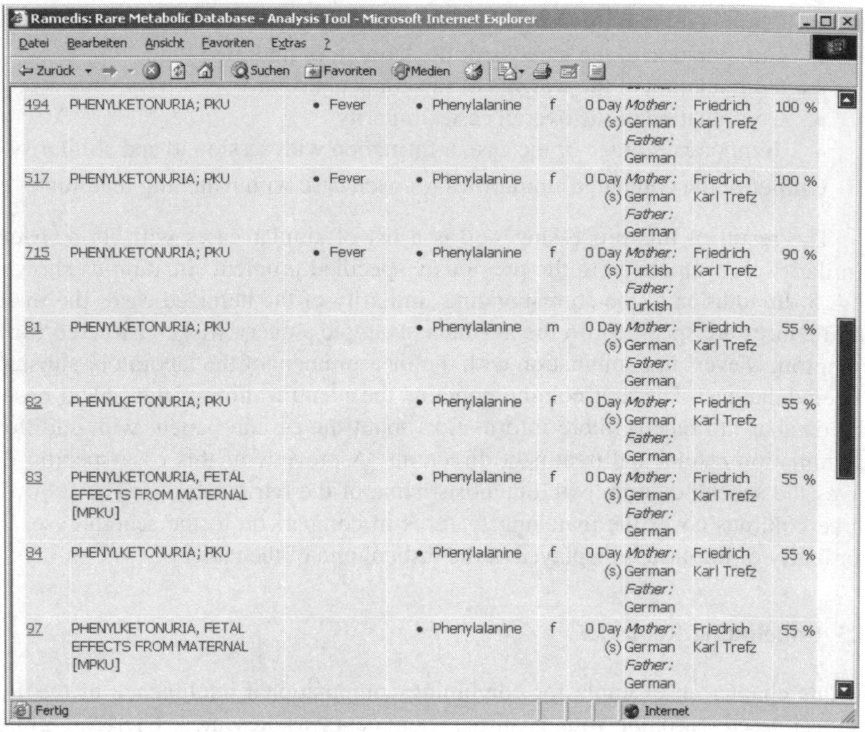

Ramedis: Rare Metabolic Database - Analysis Tool - Microsoft Internet Explorer

Datei Bearbeiten Ansicht Favoriten Extras ?

Zurück Suchen Favoriten Medien

494	PHENYLKETONURIA; PKU	• Fever	• Phenylalanine	f	0 Day (s)	*Mother:* German *Father:* German	Friedrich Karl Trefz	100 %
517	PHENYLKETONURIA; PKU	• Fever	• Phenylalanine	f	0 Day (s)	*Mother:* German *Father:* German	Friedrich Karl Trefz	100 %
715	PHENYLKETONURIA; PKU	• Fever	• Phenylalanine	f	0 Day (s)	*Mother:* Turkish *Father:* Turkish	Friedrich Karl Trefz	90 %
81	PHENYLKETONURIA; PKU		• Phenylalanine	m	0 Day (s)	*Mother:* German *Father:* German	Friedrich Karl Trefz	55 %
82	PHENYLKETONURIA; PKU		• Phenylalanine	f	0 Day (s)	*Mother:* German *Father:* German	Friedrich Karl Trefz	55 %
83	PHENYLKETONURIA, FETAL EFFECTS FROM MATERNAL [MPKU]		• Phenylalanine	f	0 Day (s)	*Mother:* German *Father:* German	Friedrich Karl Trefz	55 %
84	PHENYLKETONURIA; PKU		• Phenylalanine	f	0 Day (s)	*Mother:* German *Father:* German	Friedrich Karl Trefz	55 %
97	PHENYLKETONURIA, FETAL EFFECTS FROM MATERNAL [MPKU]		• Phenylalanine	f	0 Day (s)	*Mother:* German *Father:* German	Friedrich Karl Trefz	55 %

Fertig Internet

Fig. 5. Screenshot of the resulting list of similar cases with its respective similarity in comparison to the previously specified problem situation.

storage mechanism. For these types of diseases specific databases and information systems are available, which represent the current medical and molecular knowledge. Genes, proteins, enzymes, metabolic and signal pathways have been identified, sequenced and collected in these systems. Regarding the molecular knowledge of metabolic processes more than 500 data sources are available via the Internet. The amount of this data is increasing exponentially. Furthermore, each data source presents one specific view of the underlying molecular system. This is the reason as well as the starting point to design and develop software tools for data integration and to make that electronically available knowledge biotechnically and medical applicable. Due to this fact, an integration and analysis of the available data is essential to provide a comprehensive vision to the user. We additionally presented an approach for preparing and analyzing integrated molecular and medical data from different distributed internet data sources. By using adjusted similarity measures on special biological and medical domains like biochemical reactions a wide range of integrated data can be explored.

Acknowledgement

This work is supported by the German Ministry of Education and Research in the German Human Genome Project (Project "Modeling of gene regulatory networks for linking genotype-phenotype information") grant 01KW9962/6 and 01KW9912.

A Collection of Online Resources

In this supplementary chapter a selection of molecular biological data sources is presented briefly. A large review cannot be given in this work because the number of available databases and information systems constantly rises and for nearly each direction of biology, molecular biology and biochemistry special data sources are developed.

For more information please refer the annual appearing special edition of the Nucleic Acids Research (Baxevanis 2003) that gives a good and extensive overview of the most important representatives of the different data sources. An overview work of various genomic data source was likewise accomplished by (Borsani et al. 1998). As an application-oriented work on the use of sequence and protein databases a book of (Baxevanis and Quellette 2001) is already available in a second edition.

General Data Collections

Entrez

`http://www.ncbi.nlm.nih.gov/entrez/`

Entrez allows you to retrieve molecular biology data and bibliographic citations from the NCBI's integrated databases.

SRS

`http://srs6.ebi.ac.uk/`

The Sequence Retrieval System (SRS) is a huge systematical collection of molecular databases and analysis tools, that are uniquely formated and accessible.

Genes

CEPH

`http://www.cephb.fr/`

This is a database of genotypes for all genetic markers that have been typed in the CEPH and for reference families for linkage mapping of the human chromosomes.

EMBL

`http://www.ebi.ac.uk/`

The "EMBL Nucleotide Sequence Database" is a framed compilation of all known DNA and RNA sequences.

GDB

`http://gdbwww.dkfz-heidelberg.de/`

The GDB is the official central database for all information that is collected in the Human Genome Project.

GenBank

`http://www.ncbi.nlm.nih.gov/`

The GenBank is the NIH genetic sequence database. It is an annotated collection of all publicly available DNA sequences.

GeneCards

`http://bioinfo.weizmann.ac.il/cards/`

This database covers data of human genes, their products and their involvement in diseases.

HGMD

`http://www.uwcm.ac.uk/medicalgenetics/`

The Human Gene Mutation Database (HGMD) represents an attempt to collate known (published) gene lesions responsible for human inherited disease.

MGD

`http://www.informatics.jax.org/`

The Mouse Genome Database (MGD) contains information on mouse genetic markers, molecular segments, phenotypes, comparative mapping data etc.

MKMD

`http://research.bmn.com/mkmd/`

Mouse Knockout and Mutation Database is a regularly updated database of mouse genetic knockouts and mutations. References are directly linked to Evaluated Medline. Published by Current Biology.

MTIR

`http://www.cbil.upenn.edu/`

Data about the expression of muscle specific genes is available in this database.

PAHdb

`http://www.pahdb.mcgill.ca/`

Database that provides access to up-to-date information about mutations at the phenylalanine hydroxylase locus, including additional data.

SCPD

`http://cgsigma.cshl.org/jian`

This is a specialized promoter database of Saccharomyces cerevisiae.

VBASE

http://www.mrc-cpe.cam.ac.uk/

VBASE is a comprehensive directory of all human germline variable region sequences compiled from over a thousand published sequences.

Proteins and Enzymes

CATH

http://www.biochem.ucl.ac.uk/bsm/cath/

CATH is a hierarchical classification of protein domains. Furthermore a lexicon, which includes text and pictures describing protein class and architecture, is available.

ENZYME

http://www.expasy.org/enzyme/

ENZYME is a repository of information relative to the nomenclature of enzymes.

LIGAND

http://www.genome.ad.jp/ligand/

The Ligand Chemical Database for Enzyme Reactions (LIGAND) is designed to provide the linkage between chemical and biological aspects of life in the light of enzymatic reactions.

PDB

http://www.rcsb.org/pdb/

PDB is the single international repository for the processing and distribution of 3-D macromolecular structure data primarily determined experimentally by X-ray crystallography and NMR.

PIR

http://pir.georgetown.edu/

This database is a comprehensive, annotated, and non-redundant set of protein sequence databases in which entries are classified into family groups and where alignments of each group are available.

PMD

http://pmd.ddbj.nig.ac.jp/

The Protein Mutant Database (PMD) would be valuable as a basis of protein engineering. This database is of a type based on literature (not on proteins); that is, each entry of the database corresponds to one article which describes protein mutations.

PRF

http://www.prf.or.jp/

The Peptide Institute, Protein Research Foundation, collects information related to amino acids, peptides and proteins: Articles from scientific journals, peptide/protein sequence data, data on synthetic compounds and molecular aspects of proteins.

PRINTS

`http://www.bioinf.man.ac.uk/dbbrowser/`

PRINTS is a compendium of protein fingerprints. A fingerprint is a group of conserved motives used to characterize a protein family.

PROSITE

`http://www.expasy.ch/prosite/`

PROSITE is a database of protein families and domains. It consists of biologically significant sites, patterns and profiles that help to reliably identify to which known protein family (if any) a new sequence belongs.

REBASE

`http://rebase.neb.com/`

REBASE, the Restriction Enzyme data BASE is a collection of information about restriction enzymes, methylases, the microorganisms from which they have been isolated, recognition sequences, cleavage sites etc.

SWISS-PROT

`http://www.expasy.org/sprot/`

SWISS-PROT is a curated protein sequence database which strives to provide a high level of annotations, a minimal level of redundancy and a high level of integration with other databases.

Pathways

CSNDB

`http://geo.nihs.go.jp/`

The Cell Signaling Networks Data base (CSNDB) is a data- and knowledge- base for signaling pathways of human cells.

ExPASy

`http://www.expasy.ch/`

This database contains links to the ENZYME Database and, for each entry it also contains links to all maps of "Boehringer Biochemical Pathways" in which this entry appears.

KEGG

`http://www.genome.ad.jp/`

The Kyoto Encyclopedia of Genes and Genomes (KEGG) is an effort to computerise current knowledge of molecular and cellular biology in terms of the information pathways that consist of interacting molecules or genes.

WIT

http://wit.mcs.anl.gov/WIT2

WIT is a WWW-based system to support the curation of function assignments made to genes and the development of metabolic models.

Gene Regulation

EPD

http://www.epd.isb-sib.ch/

The Eukaryotic Promotor Database (EPD) is a collection of eukaryotic promotors in form of DNA sequences.

RegulonDB

http://www.cifn.unam.mx/

This is a database on transcriptional regulation in E. Coli.

TRANSFAC

http://www.biobase.de

This database compiles data about gene regulatory DNA sequences and protein factors binding to and acting through them.

TRRD

http://www.bionet.nsc.ru/

The Transcription Regulatory Regions Database (TRRD) is a curated database designed for accumulation of experimental data on extended regulatory regions of eukaryotic genes.

Metabolic Diseases

BIOMDB

http://www.bh4.org/

Here are collected data of mutations causing tetrahydrobiopterin deficiencies.

OMIM

http://www3.ncbi.nlm.nih.gov/

The OMIM (Online Mendelian Inheritance in Man) database is a catalog of human genes and genetic disorders authored and edited by Dr. Victor A. McKusick and his colleagues.

PATHWAY

http://oxmedinfo.jr2.ox.ac.uk/

PATHWAY is a database of inherited metabolic diseases. The database is divided into two sections: substances and diseases.

PEDBASE

http://www.icondata.com/health/pedbase/

PEDBASE is a database of pediatrics disorders. Entries are listed alphabetically by disease or condition name.

RDB

http://www.rarediseases.org/

The Rare Disease Database is a delivery system for understandable medical information to the public, including patients, families, physicians, medical institutions, and support organizations.

References

Aamodt A, Plaza E (1994) Case-Based Reasoning: Foundational Issues, Methodological Variations, and System Approaches. In AI Communications 7(1):39-59

Baxevanis AD (2003) The Molecular Biology Database Collection: 2003 update. In Nucleic Acids Research 31(1):1-12

Baxevanis AD, Quellette BFF (2001) Bioinformatics: A Practical Guide to the Analysis of Genes and Proteins. Wiley Interscience, New York

Blau N (1996) The Hyperphenyalaninemias. A Differential Diagnosis and International Database of Tetrahydrobiopterin Deficiencies. Tectum, Marburg

Borsani G, Ballabio A, Banfi S (1998) A practical guide to orient yourself in the labyrinth of genome databases. In Human Molecular Genetics 7(19):1641-1648

Collado-Vides J, Hofestädt R (2002) Gene Regulation and Metabolism - Post-Genomic Computational Approaches. MIT Press, Cambridge

Döhr S, Ehrentreich F, Frauendienst-Egger G, Hofestädt R, Hofmann O, Lange M, Mischke U, Potapov J, Scheible D, Schnee R, Scholz U, Schomburg D, Seidl K, Töpel T, Trefz FK, Werner T , Wingender E (2002) Modeling of gene regulatory networks for genotype-phenotype information. In German Human Genome Project: Progress Report 1999 - 2002, J. Wadzack, A. Haese B. Löhmer (eds.), German Human Genome Project, Managing Office of the Scientific Coordinating Committee, pp 70-71

Freier A, Hofestädt R, Lange M, Scholz U, Stephanik A (2002) BioDataServer: A SQL-based service for the online integration of life science data. In Silico Biology 2:37-57

Hofestädt R, Töpel T (2002) Case-based Support of Information Retrieval and Analysis of Molecular Data. In Proceedings of The 15th IEEE Symposium on Computer-Based Medical Systems (CBMS 2002), Maribor, Slovenia, June 4-7, 2002. pp 129-133

Kolchanov N, Hofestädt R (2003) Bioinformatics of Genome Regulation and Structure. Kluewer Academic Publishers

Kolodner JL (1992) An Introduction to Case-Based Reasoning. In Artificial Intelligence Review 6:3-34

Kromeyer-Hauschild K, Wabitsch M, Kunze D, Geller F, Geiß HC, Hesse V, von

Hippel A, Jaeger U, Johnsen D, Korte W, Menner K, Müller G, Müller JM, Niemann-Pilatus A, Remer T, Schaefer F, Wittchen HU, Zabransky S, Zellner K, Ziegler A, Hebebrand J (2001) Perzentile für den Body-mass-Index für das Kindes- und Jugendalter unter Heranziehung verschiedener deutscher Stichproben. In Monatsschrift Kinderheilkunde 149(8):807-818

Mischke U, Scholz U, Töpel T, Scheible D, Hofestädt R, Trefz FK (2001) RAMEDIS - Rare Metabolic Diseases Publishing Tool for Genotype-Phenotype Correlation. In MedInfo 2001: Proceedings of the 10th World Congress on Health and Medical Informatics, London, September 2 - 5, 2001. IOS Press, Amsterdam, pp 970-974

Porter CJ, Talbot CC, Cuticchia AJ (2000) Central Mutation Databases – A Review. In Human mutation 15:36-44

Schank RC (1982) Dynamic Memory: A Theory of Reminding and Learning in Computers and People. Cambridge University Press

Scriver CR, Hurtubise M, Konecki D, Phommarinh M, Prevost L, Erlandsen H, Stevens R, Waters PJ, Ryan S, McDonald D, Sarkissian C (2003) PAHdb 2003: What a locus-specific knowledgebase can do. In Human mutation 21:333-344

Scriver CR, Nowacki PM, Lehväslaiho H, and the Working Group (2000) Guidelines and Recommendations for Content, Structure, and Development of Mutation Databases: II Journey in Progress. In Human mutation 15:13-15

Watson I (1995) An Introduction to Case-Based Reasoning. In Progress in Case-Based Reasoning: First United Kingdom Workshop, Salford UK, January 12, 1995; Proceedings, Watson I (ed.). Springer, Berlin. pp 3-16

Prototype Based Recognition of Splice Sites

Barbara Hammer[1], Marc Strickert[1], and Thomas Villmann[2]

[1] University of Osnabrück, Department of Mathematics/Computer Science, Albrechtstraße 28, 49069 Osnabrück, Germany, {hammer,marc}@informatik.uni-osnabrueck.de

[2] University of Leipzig, Clinic for Psychotherapy and Psychosomatic Medicine, Karl-Tauchnitz-Straße 25, 04107 Leipzig, Germany, villmann@informatik.uni-leipzig.de

Summary. Splice site recognition is an important subproblem of de novo gene finding, splice junctions constituting the boundary between coding and non-coding regions in eukaryotic DNA. The availability of large amounts of sequenced DNA makes the development of fast and reliable tools for automatic identification of important functional regions of DNA necessary.

We present a prototype based pattern recognition tool trained for automatic donor and acceptor recognition. The developed classification model is very sparse and allows fast identification of splice sites. The method is compared with a recent model based on support vector machines on two publicly available data sets, a well known benchmark from the UCI-repository for human DNA [6] and a large dataset containing DNA of C.elegans. Our method shows competitive results and the achieved model is much sparser. [3]

1 Introduction

Rapid advances in biotechnology have made massive amounts of biological data available so that automated analyzing tools constitute a prerequisite to cope with huge and complex biological sequence data. Machine learning tools are used for widespread applications ranging from the identification of characteristic functional sites in genomic DNA [39], the prediction of protein secondary structure and higher structures [53], to the classification of the functionality of chemical compounds [5]. Here we will deal with a subproblem in de novo gene finding in DNA sequences of a given species, the problem of splice site recognition. For higher eukaryotic mechanisms gene finding requires the identification of the start and stop codons and the recognition of all introns, i.e. non-coding regions which are spliced out before transcription, that means all donor and acceptor sites of the sequence.

The biological splicing process is only partially understood [64]. Fig. 1 depicts a schematic view of the splicing process for eukaryotes: if genes become activated, transcription describes the process of synthesizing a copy of the coding strand of the double-stranded DNA starting at the promoter site, thereby substituting Thymine (T)

[3] The program is available at http://www.informatik.uni-osnabrueck.de/lnm/upload/

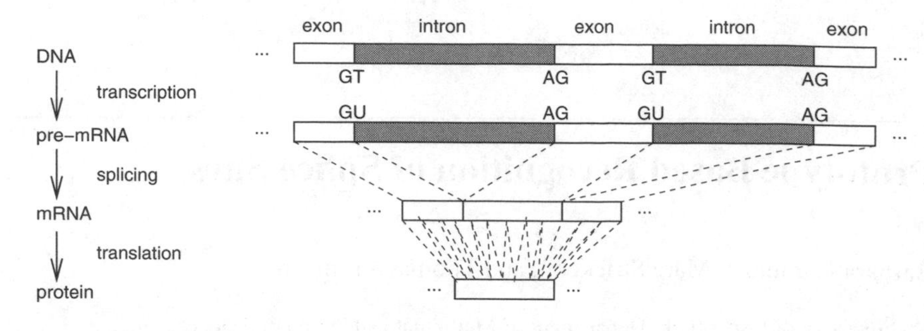

Fig. 1. Schematic view of the major steps in protein synthesis: from the point of view of computational biology, transcription consists in copying the relevant information of the DNA; splicing in eukaryotes deletes the non-coding substrings of the pre-mRNA, and translation substitutes triples of bases by new symbols for the amino acids. The resulting sequence describes the primary structure of the synthesized protein.

by the base Uracil (U). Since the information contained in the two strings is identical, DNA is often represented only by a single strand and U and T are used as synonyms in computational biology. Splicing accounts for a deletion of the non-coding regions of the pre-mRNA. The resulting mRNA is then translated into a sequence of amino acids which folds into the native state of the protein. Thereby, three letters encode one amino acid. Start and stop codons denote the boundaries of translation. For canonical splice sites, the 5' boundary or donor site of introns in mRNA usually contains the dinucleotides GT and the 3' boundary or acceptor site contains the dinucleotide AG. Non-canonical splice-sites not centered around AG and GT account for less than 1% and they are not tackled within most automated splice site recognition programs. Splice sites have strong specific consensus strings. In addition, a pyrimidine-rich region precedes AG, and a short consensus can be observed at the branch site 18-40 bp upstream of the 3' site in higher eukaryotes. Naturally, reading frames can only be found in coding regions [40]. It is not yet clear whether information beyond the pure DNA sequence such as secondary structure is also relevant for splicing [48]. In addition, splicing need not be deterministic and it might depend on external conditions. Alternative splicing is so far seldom tackled within machine learning approaches although it occurs at a much higher rate than previously expected e.g. for human DNA [17]. Alternative splicing might account for diseases, and steps towards adequate representation of alternative splicing and computational modeling are currently investigated [8, 29, 34].

Here we will consider the problem of splice site recognition based on a local window around a potential splice site. We will restrict ourselves to canonical splice sites for which large databases are avaliable for extensive evaluation of splice site recognition methods. On the one hand side, the development of reliable signal sensors for sequences based on given training data constitutes a typical problem in bioinformatics and prominent methods and typical design criteria for machine learning tools for sequences can be explained within this context. On the other hand, good splice sensors would do a nearly perfect job for ab initio gene finding. Note that a considerable amount of genes annotated in newly sequenced genomes is purely the result of computational predictions, hence the development of accurate prediction models is essential [55]. Usually, signal sensors are integrated into more general content sensors and complete gene prediction systems. However, a reduction of the number of potential splice sites and accurate splice site sensors would lead to significantly reduced computation time and more accurate overall gene prediction.

Various machine learning tools have been proposed to detect splice sites automatically based on mRNA sequences, either as stand-alone system or as subsystem within a gene-finder. Recent systems try to incorporate evolutionary patterns and combine gene location with phylogenetic or alignment information from close or weak homologues [45, 47]. However, these methods require that annotated DNA of the same or a similar family exists. Due to common patterns of splice sites, machine learning tools, which work solely on a local window around potential splice sites, constitute promising approaches, if representative data for the respective organism are available. Popular methods include feedforward neural networks [11, 42, 51, 66], statistical models [14, 31, 55, 61], logical formulas, decision trees, or combinations [12, 49, 50, 54], or, recently, support vector machines (SVM) [63]. All state-of-the-art methods achieve an accuracy above 90%. However, it is difficult to compare the models directly: results have been presented for different scenarios in the literature and training and test data are not exactly documented as discussed in [55]. It is often not documented how negative examples, i.e. local windows which do not contain splice sites, are obtained for training. They might be chosen close to true splice sites or further away. The classification accuracy highly depends on the training and test examples and a great variety in the complexity of the task for different organisms or even within one organism can be observed. Moreover, alternative evaluation criteria such as the model sensitivity and specificity are often documented for different levels which cannot directly be compared. Since it is not clear which data has been used for training, the posterior evaluation of publicly available methods, as done e.g. in [55], also gives only a partial picture.

The methods vary in complexity and accuracy. The computational expensive ones are usually more accurate, as demonstrated e.g. in the work presented in [63] where SVMs based on various kernels are proposed. Among these kernels the ones which are based on statistical models are computationally quite demanding. However, the possibility of fast classification is a prerequisite if DNA sequences are to be classified in an online e.g. Web-based environment. An additional problem consists in the fact that data are potentially high-dimensional in this task because it is not clear which size of a local window contains all relevant information for reliable splice site prediction. Too large windows, for example, might cause overfitting of the models. As a consequence, methods which guarantee good generalization independent of the input dimensionality such as SVM are particularly promising [63]. Alternatively, various models have been proposed which use small window sizes or only few data components, e.g. Boolean formulas or decision trees which use only a subset of all input dimensions [12, 50]. In addition several explicit feature selection methods have been proposed in the context of splice site recognition or other problems of bioinformatics [19, 20, 35, 41, 70]. As mentioned before, data in bioinformatics are often very high-dimensional such as sequential data, expression data derived from microarrays, or profiles, and only few training examples might be available at the same time. Dimensionality reduction is necessary to guarantee valid generalization. Besides, it might considerably speed up the classification and lead to sparse models which also allow some insight into their behavior.

Here we are interested in a splice site recognition tool which is both, accurate and sparse. For this purpose, we adapt a prototype based and similarity based learning mechanism to this special problem of splice site recognition. The algorithm yields very sparse and intuitive models by finding a small number of representative points in the space of the training examples that characterize the data. The training process can be interpreted in a mathematical precise manner as stochastic gradient descent. It shows a stable behavior also for noisy data sets. The learning objective includes an explicit term for margin optimization and good generalization ability is achieved. To avoid a critical dependence on the initialization of the algorithm, we include neighborhood cooperation of the prototypes during learning. In addition, the algorithm can be combined with any adaptive similarity measure. In particular, relevance terms weight the input dimensions, i.e. the influence of a base nucleus on the splicing, according to the contained information and therefore also high-dimensional data can be dealt with. We evaluate our learning algorithm on two data sets used in the above mentioned recent approach [63]. Both data sets are publicly available: on the one hand side, a well known (though old and possibly low quality) benchmark data set from the UCI repository [6] for human data (referred to as IPsplice) is included for which many different methods have been evaluated in comparison [6]. On the other hand a large dataset involving mRNA of C.elegans is used. For this dataset, extensive tests of various SVMs are available but no other methods have been applied so far. Since SVM showed the best performance for IPsplice as reported in [63] and constitutes one of the most successful pattern recognition tools available today, we believe that the accuracy achieved for SVM represents the state of the art and, hence, a new approach can be evaluated based on these results using the same datasets.

2 Splice site recognition methods

We start with an overview of prominent splice site recognition sensors. On the one hand side, this constitutes a representative collection of machine learning tools for classification. On the other hand it demonstrates that the problem of accurate splice sensors is not yet solved.

The basic task of splice site recognition is often tackled as a sensor problem, i.e. splice site recognition is solely performed based on the local area around potential splice sites. The overall likelihood of splice sites which also takes the length of exons and introns and further signals of the sequence into account is then computed in a second step using statistical methods and dynamic programming. For the basic problem, splice site recognition can be formulated as a classification problem: windows of fixed size around potential splice sites are chosen and constitute the input vectors. These input vectors of fixed size, whereby the nucleotides T, C, G, A can be encoded in a unary fashion in a four-dimensional vector space, are mapped to one of the classes donor, acceptor, or neither.

Different approaches have been proposed to solve this classification task for a given organism. The earliest and most simple approaches are based on consensus strings around splice sites such as PROSITE expressions [3]. This method is fast and it generalizes well to new data, however, it is not very accurate for modeling a given training set. Slightly more complex methods take the probabilities of the single nucleotides in a given training set into account. In the weight matrix model (WMM) matches of the entries with nucleotides are scored with different weightings around a potential splice site and the window size is usually chosen in the range of the consensus string. The overall score then gives the probability of splice sites [60]. This method can be interpreted as a naive Bayesian classifier [15]. Assume that $F_i(x)$ is the ith feature (i.e. nucleotide at position i) of the example x, and c is a potential class, i.e. the indication whether a splice site or not is present for x. Given x, Bayesian decision yields the class

$$c = \operatorname{argmax}_c \frac{P(F_1(x), \ldots, F_n(x)|c)P(c)}{P(F_1(x), \ldots, F_n(x))} .$$

Naive Bayes assumes independence of the features, i.e.

$$P(F_1(x), \ldots, F_n(x)) = P(F_1(x)) \cdot \ldots \cdot P(F_n(x)) .$$

These values are estimated on a given representative training set, maximizing the log likelihood for the given data. This method is fast but still not very accurate, because the assumption of independence between nucleotides cannot be maintained for DNA sequences.

Weight array matrix models generalize to first order dependencies of consecutive nucleotides by assuming a probability distribution of the form

$$P(F_1(x), \ldots, F_n(x)) = P(F_1(x))P(F_2(x)|F_1(x)) \ldots P(F_n(x)|F_{n-1}(x))$$

i.e. a first order Markov model is chosen [71]. The same model has been proposed in the articles [31, 61] within an explicit statistical context. The probabilities can efficiently be computed by means of dynamic programming, i.e. Viterbi decoding, and training can be done using standard methods such as the Baum-Welsh algorithm. This procedure is still only polynomial in time with respect to the size of the window, which is very efficient. Note that the Markov models can be embedded into more complex hidden Markov models for complete gene finding as proposed in the contributions [31, 61]. However, these models are often restricted to only first order dependencies of adjacent nucleotides.

A variety of statistical models which also take more complex dependencies into account and which are shortly mentioned below has been proposed [2, 14, 36, 61]. It is necessary to balance the complexity of the model (i.e. the window sizes and the dependencies which are taken into account) with the number of available training patterns so that the parameters can be reliably estimated. As a consequence of the comparably small data sets which were available in the past, many models take only low order dependencies into account. Expansion to higher order correlations are often accompanied by careful, possibly automated mechanisms to select only relevant information.

In HSPL a linear discriminant recognition function is based on triplet frequencies in various functional parts of the DNA and octanucleotides in potential protein coding and intron regions [61]. The method [36] proposes to also integrate a more global feature, the compositional contrast with respect to the elements U and G+C to the local windows. The approach [2] integrates pair-wise correlations of the nucleotides using a second order approximation of the probability in an appropriate expansion according to higher correlations. Since more correlations are integrated, the results can be more accurate than first order Markov models, whereby the complexity of classification is also polynomial with respect to the window size. Another extension of Markov models are Bayesian decision trees as introduced in [14]. These models can account for more general dependencies of the form

$$p(F_1(x), \ldots, F_n(x)) = p(F_1(x)) \cdot p(F_2(x)|F_1(x)) \cdot p(F_3(x)|F_1(x)) \cdot \ldots$$

where each feature might influence the probability of more than one other feature, i.e. every probability $p(F_i(x)|F_j(x))$ might occur with the restriction that all dependencies can be formulated within a directed tree structure. E.g. cycles are not allowed. The tree structure is determined automatically in the approach [14] based on pairwise mutual information of all variables. The (in an arbitrary way oriented) spanning tree which maximizes the overall mutual information between adjacent features within this tree is then chosen and the probabilities are determined by the relative frequencies of the features. Hence, appropriate global information can be detected automatically within this approach. However, the choice of the direction of the tree puts a causality assumption to processing and only limited correlations are integrated.

As an alternative, one can iteratively construct a decision tree which successively divides the splice site recognition problem into subclasses of simpler form based on particularly distinguishing features [12]. The feature which determines the respective next division is chosen using so-called maximum dependence decomposition (MDD). This computes for each pair of nucleotide positions their χ^2 statistics and chooses the position with maximum overall value. The leafs are built with simple WMM models for the remaining data. Some particularly relevant global dependencies are combined with simple WMM models. A further modification of this model has been proposed in [55]. Here MDD is combined with first order Markov models, i.e. WAMs, at the leafs. This yields very good results in comparison to several alternative methods as reported in [55]. Note that a second optimization step is integrated into the first direct prediction of potential splice sites which takes all scores in a local window into account. This global second step has been proposed in [10] and can in principle be integrated into all splice site prediction models. However, most statistical models induce severe restrictions such as limited correlation since otherwise models would become too complex and reliable parameter estimation would no longer be possible.

Alternatives to statistical models are offered by classical decision trees or induction of logical formulas e.g. in disjunctive normal form [50, 54]. Appropriate feature selection mechanisms form the core of each algorithm to avoid too complex formulas. The resulting models are usually sparse and provide explicit insight into the way of classification.

A further even more successful class of algorithms is based on more complex pattern recognition tools which can, in principle, model every possibly nonlinear dependency among variables within an appropriate window. Feedforward neural networks (FNNs) [52] constitute a very prominent approach for splice site recognition that is proposed by various authors [11, 42, 51, 66]. A complex nonlinear function is computed to approximate the decision boundaries. The functions are composed of affine functions and a fixed nonlinearity. In principle, all possible dependencies can be captured by this approach. Training takes place with some gradient descent mechanism on the training error. The resulting classifier is still fast, but usually no longer comprehensive for humans. In addition, the generalization ability of this approach scales with the number of free parameters of the model. Since the number of parameters is commonly lower bounded by the input dimension, high-dimensional data might cause problems. The approaches [11, 51] therefore include further information which allows a more direct computation of the classification borders: additional information about binding energies and explicit encoding of adjacent dinucleotides of the DNA sequence. Nevertheless, several state of the art gene finder such as NNSplice [51], NetGene2 [30], and GENIO [42] are based on neural splice site sensors.

Recently, support vector machines (SVM) have also been proposed for splice site detection [13]. Since worst case generalization bounds which are independent of the dimensionality of the input data can be derived for SVMs, they seem particularly suited for splice site recognition with potentially large window and small training sets. Informally spoken, a SVM combines a fixed nonlinear embedding of data in a possibly high-dimensional feature space with a trainable linear classifier. Training yields structural risk minimization because it chooses a separating point with largest possible margin of the data to the separation hyperplane. The nonlinear embedding is computed only implicitly using a kernel function. Appropriate design of a suitable kernel constitutes a key issue for accurate classification. The final SVM classifier can be computed as linear combination of the kernel evaluated at the respective data point and a subset of the training set which determines the output, the so-called support vectors. SVM training is polynomial with respect to the number of patterns. However, for large data sets the computational effort might still be quite high. In addition, the classification time depends on the number of support vectors which is usually given by a fraction of the training set. Hence, classification might take much longer than for alternatives such as feedforward networks or statistical models. Furthermore, the classification can often not be interpreted by humans since SVM is a black-box classifier. Another contribution [48] combines a SVM with a standard kernel with additional information related to secondary structure and binding energies leading to good results. The approach [19] introduces an input reduction scheme wrapped around SVM classification which might considerably reduce the classification time, provides further insight into the model, and might increase the accuracy. In that approach, the features which affect the prediction least are iteratively deleted. Alternative relevance determination schemes such as statistical considerations might also be appropriate [41]. Recently, several specific SVM kernels which are particularly suitable for the processing of spatio- or temporal data have been proposed, e.g. [28, 33]. The splice site recognition model proposed in [63] uses several kernels which are specifically designed for the actual problem. Kernels can be derived from statistical models such as the Fisher kernel [33] and the TOP-kernel [62]. The basic statistical model from which the kernels are derived is a hidden Markov model. The SVM approach increases the discriminative power of the original HMM model at least if enough training data are avaliable [63]. However, the complexity is quite demanding with respect to training and classification time. A third kernel applied in [63] for splice site recognition takes local correlations of time series into account and gives good results already for small training sets. Overall, SVM gives excellent results due to its inherent structural risk minimization. Still, the design of appropriate kernels is crucial, and the such designed models are usually computationally quite demanding.

All splice site recognition models can be integrated as signal sensors into complex gene finding systems and the accuracy of prediction can be increased integrating local classification results. In complex gene finding tools, often further considerations and information are integrated such as homologies to already annotated DNA from appropriate databases.

The question now occurs which method can be used for the splice site detection of a newly sequenced organism. It is complicated to directly compare the basic sensor models due to various reasons: the models have been trained for different organisms and usually the used data sets are not publicly available and mostly also not exactly documented. For example, we do not know how negative training examples (i.e. non-splice sites) have been chosen. Typical representatives for non-splice sites might vary depending on whether they are chosen in a neighborhood of splice sites or not, for example. In addition, available training data changes frequently due to newly sequenced genomes. As a consequence, reported results seldom refer to similar or even the same training sets and cannot be compared directly. An exception is a data set used in the StatLog project; competitive results for many different machine learning tools are available for this (though old and possibly outdated) data set [50]. The contribution [55] provides another comparison of several different state-of-the-art splice site predictors. This is done by testing publicly available prediction servers on the web with the same data set. It can, of courses, not be guaranteed that the used test data has not been used for training for these programs and the results might be overly optimistic. In addition, classification accuracy is measured using sensitivity and specificity with respect to donors or acceptors. Since the achieved values are different for the different programs, the results can only be compared to the newly proposed method of [55] for which a ROC curve, which relates the value $1-$specificity to the sensitivity value, is available.

Several characteristics of splice site recognition sensors (including the accuracy, of course) are interesting for the evaluation of the model efficiencies:

- Accuracy of classification: the methods should achieve a high classification accuracy. However, the accuracy might critically depend on the chosen evaluation data as discussed above. In addition, the models might differ with respect to the specificity of classification, i.e. the number of correctly predicted splice sites compared to the number of all sites which are predicted as splice sites, and the sensitivity, i.e. the number of correctly predicted splice sites compared to the number of all splice sites in the training set. The accuracy of modern systems is reported as more than 90%. More complex methods are usually more accurate and more specific to the problem. Statistical assumptions might priorly restrict the models such that perfect prediction is not possible in comparison to (in theory) approximation complete FNNs or SVMs [21, 32].
- Training effort: of course, models with low training effort are preferred to complex ones. However, training time is often less critical because it has to be done only once in an offline fashion. Since ever larger amounts of training data are available, methods which allow to further refine already trained models if new data becomes available are of particular interest. For several models such as WAMs, WMMs, and SVMs, polynomial training time can be guaranteed, whereas FNNs have worst-case exponential training time although they often perform reasonably well in practice.
- Classification effort: in particular for online systems, the classification effort should be as small as possible. Sparse models are much more efficient than more complex systems. However, most statistical models and also FNNs achieve a classification in very short time. SVMs might take much longer because their classification effort depends also on the number of support vectors, which usually scales with the training set size.
- Generalization ability and necessary training set size: data might be high--dimensional and possibly just few training data are available (or have been available in the past) so that good generalization ability of the models to unseen examples is required. Hence, statistical models are often restricted to only low order correlations although higher order correlations occur in practice. Methods for complexity reduction, such a input pruning or margin optimization in SVMs, seem particularly appropriate for these tasks.
- Sparseness of the models: sparse models often need considerably smaller classification time and possibly allow to gain insight into the classification behavior. Biological insight into relevant features and important correlations of nucleotides with respect to the classification task are, of course, interesting features of models. Here, simple statistical models are more intuitive compared to possibly complicated SVM or FNN classifiers.
- Integration of the models into a gene prediction system: usually splice sensors constitute a subtool of gene prediction tools. Therefore, it should be possible to integrate the models into larger ones. It might be necessary to vary specificity and sensitivity of the systems to account for the specific properties of the whole

prediction machinery. This can usually easily be achieved for statistical models, neural networks, and support vector machines since one can simply adapt the classification threshold in an appropriate way. Adaptation of specificity and sensitivity might, however, not be possible for computational logic approaches.

We will in the following propose a new approach, a prototype based classification. This approach achieves results which are competitive to a recent SVM approach but adding some advantages: like SVM, the prototype based classification can be interpreted as margin optimizer, hence very good generalization ability is achieved. The approach constitutes in principle a universal approximator. An explicit adaptation of the underlying metric to sequential data is possible. The approach is accompanied by a very robust and fast training method (although polynomial worst case training time cannot be guaranteed). However, unlike SVM, the model provides very sparse classifiers, the classification provides further insight into relevant factors, and moreover classification time is very low. Sensitivity and specificity of the model can be adapted as appropriate for complete gene finders.

3 Prototype based clustering

Assume that a clustering of data into C classes is to be learned and a finite training set $\{(x^i, y_i) \subset \mathbb{R}^n \times \{1, \dots, C\} \mid i = 1, \dots, m\}$ of training data is given. For splice site recognition we find 2 or 3 classes corresponding to donors, acceptors, and decoys. Denote by $X = \{x^i \mid i = 1, \dots, m\}$ all input signals of the training set. $\mathbb{R}^n$ denotes the potentially high-dimensional data space. DNA sequences are commonly encoded as concatenation of unary nucleotides. The classes are enumerated by $\{1, \dots, C\}$. We denote the components of a vector $x \in \mathbb{R}^n$ by subscripts, i.e., $x = (x_1, \dots, x_n)$. Learning vector quantization (LVQ) as introduced by Kohonen combines the accuracy of supervised training methods with the elegance and simplicity of self-organizing systems [37]. LVQ represents every class c by a set $W(c)$ of weight vectors (prototypes) in $\mathbb{R}^n$. Weight vectors are denoted by w^r and their respective class label is referred to by c_r. A new signal $x \in \mathbb{R}^n$ is classified by the winner-takes-all-rule by an LVQ network, i.e.

$$x \mapsto c(x) = c_r \text{ such that } d(x, w^r) \text{ is minimum} .$$

$d(x, w^r) = \|x - w^r\|^2 = \sum_i (x_i - w_i^r)^2$ denotes the squared Euclidean distance of the data point x to the prototype w^r. The respective closest prototype w^r is called winner or best matching unit. The subset

$$\Omega_r = \{x^i \in X \mid d(x^i, w^r) \text{ is minimum}\}$$

is called receptive field of neuron w^r.

The training algorithm of LVQ aims at minimizing the classification error on the given training set. I.e., the difference of the points belonging to the cth class, $\{x^i \in X \mid y_i = c\}$, and the receptive fields of the corresponding prototypes,

$\bigcup_{w^r \in W(c)} \Omega_r$, is minimized by the adaptation process. Training iteratively presents randomly chosen data from the training set and adapts the respective closest prototype by Hebbian learning. Hebbian learning refers to the intuitive scheme to emphasize already correct classifications and to de-emphasize bad ones. If a vector x^i is presented, the update rule for the winner w^r has the form

$$\triangle w^r = \begin{cases} \epsilon \cdot (x^i - w^r) & \text{if } c^r = c(x^i) \\ -\epsilon \cdot (x^i - w^r) & \text{otherwise} \end{cases}$$

$\epsilon \in (0,1)$ is an appropriate learning rate. As explained in [56] this update can be interpreted as a stochastic gradient descent on the cost function

$$\text{Cost}_{\text{LVQ}} = \sum_{x^i \in X} f_{\text{LVQ}}(d_{r^+}, d_{r^-}) \,.$$

d_{r^+} denotes the squared Euclidean distance of x^i to the closest prototype w^{r^+} labeled with $c_{r^+} = y_i$ and d_{r^-} denotes the squared Euclidean distance to the closest prototype w^{r^-} labeled with a label c_{r^-} different from y_i. For standard LVQ, the function is

$$f_{\text{LVQ}}(d_{r^+}, d_{r^-}) = \begin{cases} d_{r^+} & \text{if } d_{r^+} \le d_{r^-} \\ -d_{r^-} & \text{otherwise} \end{cases}$$

Obviously, this cost function is highly discontinuous and instabilities arise for overlapping data distributions. Various alternatives have been proposed to achieve a more stable behavior of training also in case of overlapping classes or noisy data. LVQ2.1 as proposed by Kohonen optimizes a different cost function which is obtained by setting in the above sum $f_{\text{LVQ2.1}}(d_{r^+}, d_{r^-}) = I_w(d_{r^+} - d_{r^-})$ whereby I_w yields the identity within a window in which adaptation of LVQ2.1 takes place, and I_w vanishes outside. Still this choice might produce an instable dynamic, i.e. prototypes might diverge due to the fact that repelling forces from the term d_{r^-} might be larger than attracting forces from the term d_{r^+}. To prevent this behavior as far as possible, the window within which adaptation takes place must be chosen carefully.

Note that LVQ2.1 explicitly optimizes the term $d_{r^+} - d_{r^-}$. It has been shown recently that LVQ can therefore be interpreted as structural risk minimization scheme: according to [18], the related term

$$(\|x^i - w^{r^-}\| - \|x^i - w^{r^+}\|)/2$$

constitutes the so-called hypothesis margin of the classifier. The hypothesis margin denotes the distance that the classifier can be changed (in an appropriate norm) without changing the classification. As proved in [18], the generalization ability of a LVQ-classifier can be estimated based on the hypothesis margin independently of the dimensionality of the given input data. Hence LVQ can be interpreted as a large margin optimizer similar to SVMs thus aiming at optimizing the generalization ability during training, which is a valuable feature.

An alternative to LVQ2.1 which also explicitly optimizes the hypothesis margin has been proposed in [56], so-called generalized LVQ (GLVQ). The respective cost function can be obtained by setting

$$f_{\mathrm{GLVQ}}(d_{r^+}, d_{r^-}) = \mathrm{sgd}\left(\frac{d_{r^+} - d_{r^-}}{d_{r^+} + d_{r^-}}\right)$$

whereby $\mathrm{sgd}(x) = (1 + \exp(-x))^{-1}$ denotes the logistic function. As discussed in [57], the additional scaling factors avoid numerical instabilities and divergent behavior also for overlapping data distributions. Note that the term $(d_{r^+} - d_{r^-})/(d_{r^+} + d_{r^-})$ evaluates the classification of LVQ with a number between -1 (correct classification with large margin) and 1 (entirely wrong classification). The default classification threshold is given by 0. The additional scaling caused by the logistic function further emphasizes the relevant range of values around 0. Hence, the data points for which the closest correct and the closest wrong prototype have almost the same distance also contribute to the learning. Update rules for GLVQ can be obtained by taking the derivatives with respect to the prototypes:

$$\triangle w_{r^+} = \epsilon^+ \cdot \mathrm{sgd}'_{\mu(x^i)} \cdot \xi^+ \cdot 2 \cdot (w^{r^+} - x^i)$$

and

$$\triangle w_{r^-} = -\epsilon^- \cdot \mathrm{sgd}'_{\mu(x^i)} \cdot \xi^- \cdot 2 \cdot (w^{r^-} - x^i)$$

denotes the update of the closest correct and closest wrong prototype, given an example x^i, whereby ϵ^+ and $\epsilon^- \in (0,1)$ are the learning rates, the logistic function is evaluated at position $\mu(x^i) = (d_{r^+} - d_{r^-})/(d_{r^+} + d_{r^-})$, and

$$\xi^+ = \frac{2 \cdot d_{r^-}}{(d_{r^+} + d_{r^-})^2}$$

and

$$\xi^- = \frac{2 \cdot d_{r^+}}{(d_{r^+} + d_{r^-})^2}$$

denote the derivatives of $f_{\mathrm{GLVQ}}(d_{r^+}, d_{r^-})$ with respect to d_{r^+} and d_{r^-}.

Alternative modifications of LVQ with cost function optimization have been proposed, such as the recent statistic interpretation developed in [59]. However, the proposed cost functions do often no longer optimize the hypothesis margin leading to a worse generalization ability compared to GLVQ.

Since GLVQ is an iterative update scheme, it might suffer from the problem of local optima: it constitutes a stochastic gradient descent on a possibly highly multimodal cost function and the result might severely depend on the initialization of prototypes. This problem already occurs for simple LVQ and LVQ2.1, Kohonen proposes one possible solution: one can include neighborhood cooperation of the prototypes to ensure a faithful representation of all modes of the given data [38]. Kohonen proposes to combine LVQ with the so-called self-organizing map, a very intuitive and well established unsupervised learning scheme which allows to find a topological representation of given data points by a lattice of neurons [37]. The fact that GLVQ can be interpreted as a stochastic gradient descent offers a particularly efficient alternative way of neighborhood integration. We can combine the cost function of GLVQ with the unsupervised Neural Gas algorithm (NG), the latter offering a particularly robust and efficient unsupervised learning scheme to spread prototypes faithfully among a given data set [43, 44]. Unlike the self-organizing map, NG does not rely on a fixed priorly chosen neighborhood structure but chooses automatically a data-optimum lattice and computational artefacts and topological mismatches do not occur for NG [68]. Given prototypes w^r from a set W and data points x^i (both without class labels), NG optimizes the cost function

$$\text{Cost}_{\text{NG}} = \frac{1}{C(\gamma, K)} \sum_{w^r \in W} \sum_{x^i \in X} h_\gamma(r, x^i, W) d(x^i, w^r)$$

where

$$h_\gamma(r, x^i, W) = \exp\left(-\frac{k_r(x^i, W)}{\gamma}\right)$$

denotes the degree of neighborhood cooperativity, $k_r(x^i, W)$ yields the rank of w^r which is the number of prototypes w^p for which $d(x^i, w^p) < d(x^i, w^r)$ is valid, and $C(\gamma, K)$ is a normalization constant depending on the neighborhood range γ and cardinality K of W. The learning rule is given by

$$\triangle w^r = \epsilon \cdot h_\gamma(r, x^i, W)(x^i - w^r)$$

where $\epsilon > 0$ is the learning rate.

Supervised Neural Gas

As already proposed in [25], NG can directly be combined with GRLVQ, integrating neighborhood cooperation for prototypes of the same class, thus avoiding local optima and accelerating convergence. The cost function of this supervised neural gas (SNG) algorithm is

$$E_{\text{SNG}} = \sum_{x^i \in X} \sum_{w^r \in W(y^i)} \frac{h_\gamma(r, x^i, W(y^i)) \cdot f_{\text{SNG}}(d_r, d_{r^-})}{C(\gamma, K(y^i))}$$

whereby

$$f_{\text{SNG}}(d_r, d_{r^-}) = f_{\text{GLVQ}}(d_r, d_{r^-}) = \text{sgd}\left(\frac{d_r - d_{r^-}}{d_r + d_{r^-}}\right)$$

and d_r denotes the squared Euclidean distance of x^i to w^r. $K(y^i)$ denotes the cardinality of the set of prototypes labeled by y^i, i.e. $|W(y^i)|$. w^{r^-} denotes the closest prototype not in $W(y^i)$. Due to the NG-dynamics, all prototypes of a specific class are adapted towards the given data point, preventing neurons from being idle or repelled from their class. The simultaneous GLVQ dynamics makes sure that those class borders are found which yield a good classification. In addition, the cost function includes terms related to the hypothesis margin, just like GLVQ and original LVQ. Note that vanishing neighborhood cooperativity $\gamma \to 0$ yields the original cost function of GLVQ.

The update formulas for the prototypes are obtained by taking the derivative. For each x^i, all prototypes $w^r \in W(y^i)$ are adapted by

$$\triangle w^r = \epsilon^+ \cdot \frac{\mathrm{sgd}'|_{\mu^r(x^i)} \cdot \xi_r^+ \cdot h_\gamma(r, x^i, W(y^i))}{C(\gamma, K(y^i))} \cdot 2 \cdot (x^i - w^r)$$

and the closest wrong prototype is adapted by

$$\triangle w^{r^-} = -\epsilon^- \sum_{w^r \in W(y^i)} \frac{\mathrm{sgd}'|_{\mu^r(x^i)} \cdot \xi_r^- \cdot h_\gamma(r, x^i, W(y^i))}{C(\gamma, K(y^i))} \cdot 2 \cdot (x^i - w^{r^-})\,.$$

ϵ^+ and $\epsilon^- \in (0,1)$ are learning rates and the derivative of the logistic function is evaluated at position

$$\mu^r(x^i) = \frac{d_r - d_{r^-}}{d_r + d_{r^-}}\,.$$

The terms ξ are again obtained as derivative of f_{SNG} as

$$\xi_r^+ = \frac{2 \cdot d_{r^-}}{(d_r + d_{r^-})^2}$$

and

$$\xi_r^- = \frac{2 \cdot d_r}{(d_r + d_{r^-})^2}\,.$$

The precise derivation of these formulas also for continuous data distributions can be found in [24]. Note that the original updates of GLVQ are recovered if $\gamma \to 0$. For positive neighborhood cooperation, all correct prototypes are adapted according to a given data point such that also neurons outside their class become active. Eventually, neurons become spread among the data points of their respective class. Since all prototypes have a repelling function on the closest incorrect prototype, it is advisable to choose ϵ^- one magnitude smaller than ϵ^+. As demonstrated in [25] also highly multimodal classification tasks can be solved with this modification of GLVQ. Note that GLVQ allows insight into the classification behavior by looking at the prototypes. Often, a comparably small number of prototypes already yields good classification accuracy.

4 Relevance learning and general similarity measures

SNG does no longer critically depend on initialization and shows stable behavior also for overlapping classes. However, it depends crucially on the underlying Euclidean metric. It is based on the assumption that data form compact clusters in Euclidean space. This assumption is not met if heterogeneous data are dealt with for which the input dimensions have different relevance for the classification. The situation is particularly critical if high-dimensional data are considered, where a large number of possibly less relevant and noisy dimensions disturbs the whole classification. As discussed beforehand, biological data are often high-dimensional and heterogeneous. Feature selection mechanisms have been proposed in various different scenarios [19, 20, 35, 41, 70]. In the case of splice site recognition, we would like to use large windows to make sure that all relevant information is contained. It can be expected that nucleotides close to potential splice sites within regions of consensus strings are more relevant for classification than other entries. In addition, local correlations might play an important role as indicated by various models based on correlations of adjacent dinucleotides such as WAMs.

SNG possesses a cost function and it is straightforward to substitute the Euclidean metric by an alternative, possibly adaptive one. A particularly efficient and powerful model has been introduced in [26]: the Euclidean metric is substituted by a weighted Euclidean metric which incorporates relevance factors for the input dimensions. I.e., $d(x,y) = \|x-y\|^2$ is substituted by

$$d^\lambda(x,y) = \sum_i \lambda_i (x_i - y_i)^2$$

whereby $\lambda_i \geq 0$ and $\sum_i \lambda_i = 1$. Appropriate scaling terms λ_i allow to choose small relevance factors for less relevant or noisy dimensions such that high-dimensional or heterogeneous data can be accounted for with an appropriate choice of λ_i. Since appropriate relevances are not a priori known, we adapt also the relevance terms with a stochastic gradient descent on the cost function. The cost function of supervised relevance neural gas (SRNG), i.e. SNG with weighted metric, is

$$E_{\text{SRNG}} = \sum_{x^i \in X} \sum_{w^r \in W(y_i)} \frac{h_\gamma(r, x^i, W(y_i)) \cdot f_{\text{SNG}}(d_r^\lambda, d_{r^-}^\lambda)}{C(\gamma, K(y_i))} \tag{1}$$

whereby $d_r^\lambda := d^\lambda(x^i, w_r)$, r^- denotes the closest prototype which does not belong to $W(y_i)$ measured according to the similarity measure d^λ, and the rank of prototypes is computed using similarity measure d^λ, too. The updates with respect to the prototypes can be obtained as beforehand. Given x^i, all prototypes $w^r \in W(y^i)$ are adapted by

$$\triangle w^r = -\epsilon^+ \cdot \frac{\text{sgd}'|_{\mu^r(x^i)} \cdot \xi_r^+ \cdot h_\gamma(r, x^i, W(y^i))}{C(\gamma, K(y^i))} \cdot \frac{\partial d_r^\lambda}{\partial w^r} \tag{2}$$

and the closest wrong prototype is adapted by

$$\triangle w'^- = \epsilon^- \cdot \sum_{w^r \in W(y^i)} \frac{\text{sgd}'|_{\mu^r(x^i)} \cdot \xi_r^- \cdot h_\gamma(r, x^i, W(y^i))}{C(\gamma, K(y^i))} \cdot \frac{\partial d_{r^-}^\lambda}{\partial w^{r^-}} \tag{3}$$

whereby the logistic function is evaluated at position

$$\mu^r(x^i) = \frac{d_r^\lambda - d_{r^-}^\lambda}{d_r^\lambda + d_{r^-}^\lambda}.$$

The terms ξ are again obtained as derivative of f_{SNG} as

$$\xi_r^+ = \frac{2 \cdot d_{r-}^\lambda}{(d_r^\lambda + d_{r-}^\lambda)^2} \quad \text{and} \quad \xi_r^- = \frac{2 \cdot d_r^\lambda}{(d_r^\lambda + d_{r-}^\lambda)^2} .$$

For the squared Euclidean metric, we obtain

$$\frac{\partial d_r^\lambda}{\partial w^r} = -2 \cdot \Lambda \cdot (x^i - w^r)$$

whereby Λ denotes the matrix with diagonal elements λ_i. The updates of the relevance factors are given by

$$\Delta\lambda = \epsilon^\lambda \cdot \sum_{w^r \in W(y^i)} \frac{\text{sgd}'|_{\mu^r(x^i)} \cdot h_\gamma(r, x^i, W(y^i))}{C(\gamma, K(y^i))} \tag{4}$$

$$\cdot \left(\xi_r^+ \cdot \frac{\partial d_r^\lambda}{\partial \lambda} - \xi_r^- \cdot \frac{\partial d_{r-}^\lambda}{\partial \lambda} \right) . \tag{5}$$

Here $\epsilon^\lambda \in (0,1)$ constitutes the learning rate for λ. It should be chosen some magnitudes smaller than the learning rates for the prototypes such that adaptation of the prototypes takes place in a quasi stationary environment and convergence and generalization is optimized for each choice of relevance terms. For the squared Euclidean metric, it holds

$$\frac{\partial d_r^\lambda}{\partial \lambda_l} = (x_l^i - w_l^r)^2 .$$

After each adaptation step, the relevance terms must be normalized to ensure the constraint $\lambda_i \geq 0$ and $\sum_i \lambda_i = 1$. The precise derivation of these formulas also for continuous data distributions can again be found in [24].

Note that input features with different relevances are now automatically taken into account. In addition the resulting relevance profile of training allows to gain insight into the problem since the resulting relevance profile indicates which input features contribute most to the classification. The squared Euclidean metric constitutes one very efficient possible distance measure to compare prototypes and data points. Note, however, that it is in principle possible to use an arbitrary differentiable similarity measure $d^\lambda(x^i, w^r)$ including possibly adaptive terms λ. The similarity measure should be some positive function which measures in an adequate way the closeness of prototypes and data points. It is well known that alternatives to the Euclidean metric might be better suited for non Gaussian data e.g. because of larger robustness with respect to noise or other properties [16]. In our experiments, we use two further similarity measures which show more robust behavior and increase the classification accuracy.

The weighted quartic metric

$$d_4^\lambda(x, y) := \sum_i \lambda_i^2 (x_i - y_i)^4 \tag{6}$$

leads to faster convergence and better classification accuracy in some cases. Note that this measure punishes large deviations of the data point from the prototype more than smaller ones in comparison to the weighted Euclidean distance. It seems particularly suited if the entries are not Gaussian but more sharply clustered. The quartic similarity then better matches the test whether a large number of entries can be found within the sharp clusters provided by the prototype coefficients.

In analogy to the so-called locality improved kernel (LIK) which has been proposed for splice site recognition to take local correlations of spatial data into account [62, 63], we use the following similarity measure: assume data points have the form $x = (x_1, \dots, x_n)$ and local correlations of neighbored entries x_i, x_{i+1} might be relevant for the similarity; x might, for example, represent a local window of length n around a potential splice site. d_L then computes

$$d_L^\lambda(x, y) := \sum_{i=1}^{n} \lambda_i s_i(x, y) \tag{7}$$

whereby

$$s_i(x, y) = \left(\sum_{j=-l}^{l} \frac{b_j}{b_{\text{norm}}} (x_{i+j} - y_{i+j})^2 \right)^\beta$$

measures the correlation of the distances of the entries within a local window around position i of the two data points. $\beta > 0$ is typically chosen as a small natural number, b_j is a factor which is decreasing towards the borders of the local window such as $b_j = 1/(\delta \cdot |j| + 1)$ with $\delta > 0$, $b_{\text{norm}} = \sum_{j=-l}^{l} b_j$, and l denotes the radius of the local windows. At the borders of the range of indices of x and y, adaptation of the window size is necessary. $\lambda_i \geq 0$ are adaptive values with $\sum_i \lambda_i = 1$. In the case of DNA-sequences, entries consist of nucleotides which are embedded in a low-dimensional vector space, i.e. $x_i = (x_{i,1}, \dots, x_{i,N}) \in \mathbb{R}^N$ for some small N (e.g. $N = 4$ for unary encoding of nucleotides), and neighboring vectors x_i should be compared. In this case we can generalize

$$s_i(x, y) = \left(\sum_{j=-l}^{l} \frac{b_j}{b_{\text{norm}}} \sum_{k=1}^{N} (x_{i+j,N} - y_{i+j,N})^2 \right)^\beta$$

if x and y contain entries which are vectors in $\mathbb{R}^N$. Research reported in [51, 61, 63] indicates that classification results for splice site recognition can be improved if local correlations of nucleotides are taken into account. This is quite reasonable since local correlations can easily detect, on the one hand side, reading frames and, on the other hand, local consensus strings. LIK seems particularly suited for the task of splice site recognition.

Note that we can easily substitute the weighted Euclidean metric by the quartic similarity measure or the locality improved measure by substituting the metric in the cost function 1 by the terms as defined in equation 6 or equation 7. The learning rules are obtained from equations 2, 3, and 4 using the equalities

$$\frac{\partial d_4^\lambda(x, y)}{\partial x} = 4 \cdot \lambda_i^2 \cdot (x_i - y_i)^3, \quad \frac{\partial d_4^\lambda(x, y)}{\partial y} = -4 \cdot \lambda_i^2 \cdot (x_i - y_i)^3,$$

and

$$\frac{\partial d_4^\lambda(x,y)}{\partial \lambda_i} = 2 \cdot \lambda_i \cdot (x_i - y_i)^4$$

for the quartic similarity measure and

$$\frac{\partial d_L^\lambda(x,y)}{\partial x_i} = \sum_{p=i-l}^{i+l} 2 \cdot \lambda_p \cdot \beta \cdot (s_p(x,y))^{\frac{\beta-1}{\beta}} \cdot \frac{b_{i-p}}{b_{\text{norm}}} \cdot (x_i - y_i)\,,$$

$$\frac{\partial d_L^\lambda(x,y)}{\partial y_i} = -\sum_{p=i-l}^{i+l} 2 \cdot \lambda_p \cdot \beta \cdot (s_p(x,y))^{\frac{\beta-1}{\beta}} \cdot \frac{b_{i-p}}{b_{\text{norm}}} \cdot (x_i - y_i)\,,$$

and

$$\frac{\partial d_L^\lambda(x,y)}{\partial \lambda_i} = s_i(x,y)\,.$$

5 A note on the generalization ability

The generalization ability of a classifier refers to the expected error for data which have not been used for training in comparison to the training error. There exist various ways to formalize and prove the generalization ability of classifiers such as the popular VC-theory [67]. Since data are often high-dimensional in problems of bioinformatics requiring many parameters to be tuned during training, methods which directly optimize the generalization ability of the classifier seem particularly suited for these tasks. One of the most prominent methods which directly aims at structural risk minimization is the SVM. LVQ constitutes an alternative for which dimension independent generalization bounds have recently been derived [18]. Here, we discuss which aspects can be transferred to SRNG. We mention two possible argumentation lines, one for fixed similarity measure and one which also captures adaptive relevance terms.

As already mentioned, generalization bounds for prototype based classifiers in terms of the so-called hypothesis margin have been derived in [18]. The hypothesis margin is given by the distance of the closest correct prototype compared to the distance of the closest incorrect prototype. GLVQ as well as LVQ2.1 directly aim at margin optimization and excellent generalization can thus be expected also for high-dimensional data. We have substituted the Euclidean metric by different similarity measures and the above argument does no longer apply. However, the bounds transfer to a general similarity if the situation can be interpreted as a kernelized version of the original algorithm. Kernelization constitutes a popular trick of which the SVM constitutes the most prominent application. In our case, a kernel refers to a function $\Phi : \mathbb{R}^n \to X$ from the data space $\mathbb{R}^n$ to some Hilbert space X such that the similarity measure d of our algorithm can be written as

$$d(x,y) = \|\Phi(x) - \Phi(y)\|$$

whereby d refers to the respective similarity which we use in the actual version of SRNG, and $\|\cdot\|$ refers to the metric of the Hilbert space X. It has been shown in [58] that such Φ can be found if d constitutes a real-valued symmetric function with $d(x,x) = 0$ such that $-d$ is conditionally positive definite, i.e. for all $N \in \mathbb{N}$, $c_1, \ldots, c_N \in \mathbb{R}$ with $\sum_i c_i = 0$ and $x^1, \ldots, x^N \in \mathbb{R}^n$ the inequality $\sum_{i,j} c_i c_j \cdot (-1) \cdot d(x^i, x^j) \geq 0$ holds. As an example, functions of the form $\|\mathbf{x}-\mathbf{y}\|^\beta$ for an arbitrary metric $\|\cdot\|$ and $\beta \in (0,2]$ fulfill these properties. The similarity measures which we use, the quartic measure and LIK, have been derived in analogy to well-known kernels of SVM, but they do unfortunately not fulfill this condition. Nevertheless, we could observe excellent generalization ability in our experiments.

Another more fundamental problem which prohibits the direct application of these generalization bounds for SRNG consists in the fact that we change the similarity measure during training: the relevance factors are adapted according to the given training data. If arbitrary changes of the kernel were allowed, no generalization bounds would hold [27]. If adaptation is restricted for example to convex combinations of fixed kernels, bounds have been established [9]. Arguments along this line could also be applied to SRNG. However, another more direct argumentation is presented in [23] which derives a large margin generalization bound for the basic weighted squared similarity measure also for adaptive relevance terms. We sketch this result in the remaining part of this section.

Assume that a SRNG network with the weighted squared Euclidian similarity measure and adaptive relevance terms is trained on m data points (x^1, y^1), ..., $(x^m, y^m) \in \mathbb{R}^n \times \{-1, 1\}$ for a two-class problem with inputs x^i and outputs y^i. The two classes are referred to by -1 and 1, for simplicity. We assume that the data points are chosen randomly according to a fixed unknown probability distribution P on $\mathbb{R}^n \times \{-1, 1\}$. Given some example x, denote by d_{r+} the closest prototype labeled with 1 and by d_{r-} the closest prototype labeled with -1. A point x with desired output y is classified correctly iff

$$y \cdot (-d_{r+} + d_{r-}) \geq 0$$

The quantity

$$\hat{E}_m := \sum_{i=1}^{m} H\left(y^i \cdot (-d_{r+} + d_{r-})\right) / m$$

thus measures the percentage of misclassifications on the given training set whereby $H : \mathbb{R} \to \{0, 1\}$ denotes the standard Heaviside function with $H(t) = 0 \iff t < 0$. The term generalization refers to the fact that the number of misclassifications on the training set $\hat{E}_m$ (which is usually small since it is optimized during training) is representative for the number of misclassifications for new data points x which have not been used for training. In mathematical terms, generalization refers to the fact that $\hat{E}_m$ approximates the expected value of $y \cdot (-d_{r+} + d_{r-})$ if (x, y) is chosen randomly according P. We are interested in guarantees for this property. Note, however, that the empirical error $\hat{E}_m$ as defined above does not take a large margin into account. The margin is thereby given by the value $|-d_{r+} + d_{r-}|$, i.e. the difference between the distances to the closest correct and incorrect prototypes. This is related to the hypothesis margin as introduced in [18] and it is directly optimized during training due to the cost function of SRNG. We are interested in the generalization ability of classifiers with margin at least ρ for some $\rho > 0$. Following the approach [4], we therefore introduce the following loss function

$$L : \mathbb{R} \to \mathbb{R}, t \mapsto \begin{cases} 1 & \text{if } t \leq 0 \\ 1 - t/\rho & \text{if } 0 < t \leq \rho \\ 0 & \text{otherwise} \end{cases}$$

The term

$$\hat{E}^L_m := \sum_{i=1}^{m} L(y^i \cdot (-d_{r+} + d_{r-}))/m$$

collects all misclassifications of the classifier as beforehand, and it also punishes correctly classified points which have a small margin. We refer to this term as the empirical error with margin ρ in the following. Generalization with margin ρ refers to the fact that $\hat{E}^L_m$ is representative for the expectation of the term $y \cdot (-d_{r+} + d_{r-})$ for randomly chosen (x, y) according to P, i.e. the expected percentage of misclassifications for new data points. It has been shown in [23] that the expected percentage of misclassifications can be upper bounded with probability $1 - \delta$ by $\hat{E}^L_m$ and a term of order

$$\frac{p^2}{\rho \cdot \sqrt{m}} \cdot \left(b^3 + \sqrt{\ln 1/\delta} \right)$$

whereby p denotes the number of prototypes of the SRNG network and b denotes some constant which limits the length of each training point x^i and the length of each prototype vector w. Note that this bound does not depend on the dimensionality of the data and it scales inversely to the margin parameter ρ. Thus it gives a direct large margin bound for SRNG networks with weighted squared similarity measure and adaptive relevance terms.

6 Experiments

We will deal with two data sets which are publicly available at the web and for which results have been published for alternative methods. The IPsplice dataset consists of 765 acceptors, 767 donors, and 1654 decoys from human DNA [6]. The problem can thus be tackled as a classification problem with three classes, acceptors, donors, and neither. The size of the window is fixed to 60 nucleotides centered around potential splice sites in the dataset. It is not documented of how decoys have been collected. The data set is rather old. However, as a well known benchmark which has been used in the StatLog project competitive results are available for many different machine learning approaches to which we can compare our method.

The (current) C.elegans data consists of training data sets including 1000 and 10000 examples (five different data sets for each size) for training and a set of 10000 data points for testing. The data sets are available in the Internet [63]. They have been derived from the Caenorhabditis elegans genome [1]. We will only consider the task of separating acceptors and decoys, i.e. a two-class problem is dealt with. Classification results for these sets have been published for SVMs with various different kernels and for hidden Markov models in [63]. Data is presented within a fixed window size of 50 nucleotides. The decoys have been collected at positions close to the true acceptor sites and centered around AG. Approximately twice as many decoys as acceptors are present.

In all experiments, the 4 nucleotides are encoded as tetrahedron points in a 3-dimensional vector space. Prototypes are always initialized with small entries, and the relevance factors of the metrics are all equal at the beginning of training.

IPSplice

For IPSplice, we trained SRNG with 8 prototypes for each class (acceptor, donor, and neither). We used the weighted Euclidean metric d^2_λ and the locality improved kernel with local window with radius $l = 3$ such that local correlations, in particular also possible reading frames are taken into account. The exponent is chosen as $\beta = 3$, the weighting within the windows as $\delta = 0.5$ for LIK. Also all further training parameters have been optimized for this task. We choose the learning rate for the prototypes $\epsilon^+ = 0.004$, ϵ^- is adapted from a small value to $0.075 \cdot \epsilon^+$ (for the weighted Euclidean metric) and $0.5 \cdot \epsilon^+$ (for LIK) during training. The learning rate for the relevance terms is $\epsilon^\lambda = 10^{-8}$. The initial neighborhood range is chosen as 8 and multiplicatively decreased after each training epoch by 0.9999

Training converged after 2000 epochs and training patterns are presented in random order. We achieve in a 10-fold crossvalidation with randomly chosen training set including 2000 data points and test set including the remaining 1186 data points the test set accuracy $95.627\% \pm 0.536\%$ for the weighted Euclidean similarity and $96.458\% \pm 0.27\%$ for LIK. Results for several alternative machine learning tools are available as indicated in Table 1. (taken from [46, 50, 63]). The methods include a decision tree approach (C4.5), a nearest neighbor classifier (k-NN), a neural network (BP), a radial basis function network (RBF), two statistical models (DISC and HMM), and three SVM approaches. The SVMs use kernels which have been explicitly designed for use in splice site recognition: the locality improved kernel for SVM is comparable to the similarity measure LIK which we used in our experiment. In addition, two kernels which have been derived from statistical models and thus combine well established statistical modeling with the additional discriminative power of SVMs are used, referred to as Fisher kernel (FK) and TOP kernel. The results for the SVM and HMM correspond to the total error rate as reported in [63]. The other results are taken from the StatLog project in which a large number of machine learning tools has been systematically compared [46]. Thereby, radial basis functions performed best for all methods tested in the Statlog project in this task. Their storage requirement is thereby comparably large (a factor of about 8 compared to standard feedforward networks) since a large number of centres has to be taken into account. Obviously, the SVM in combination with LIK can improve the results achieved with the methods tested in the StatLog project.

$\text{our}_{L_2^\lambda}$	our_{LIK}	SVM_{LIK}	SVM_{TOP}	SVM_{FK}	HMM	DISC	k-NN	C4.5	BP	RBF
95.6	96.5	96.3	94.6	94.7	94	94.1	86.4	92.4	91.2	95.9

Table 1. Accuracy (in %) of various methods achieved on the IPsplice dataset, the classification accuracy on the test set is reported for the models. The results for SVM and HMM correspond to the total error rates of binary classifiers [63].

Our result for the weighted Euclidean similarity measure is already comparable to the best results which have been achieved in [63, 46]. Our results for LIK are even slightly better, whereby the SVM provides state of the art results, since it constitute very powerful methods also for high-dimensional data sets with good generalization ability and the possibility to also include higher order correlations of nucleotides. The combination of statistical models and SVMs combines the advantages of both, however, shows slightly worse results possibly due to overfitting since the data set is comparably small. Our models show the same good or even better results and demonstrate excellent generalization ability. Note that standard feedforward networks (BP) perform quite badly although they are as powerful as SVMs. This is due to the comparably high-dimensional inputs which cause strong overfitting of the network. The combination of powerful models with explicit mechanisms to cope with high-dimensional input data are particularly useful.

Note that our models only use 8 prototypes per class, i.e. they can be described with 24 vectors. Our models are sparse and the classification is very fast. Classification complexity using HMMs depends on the number of interior states, which is chosen as about 10 in the approach [63]. Hence the classification time of HMMs is competitive to our model. SVMs can be expressed in terms of the support vectors which constitute a fraction of the data set. For the IPSplice data set, an order of roughly 100 support vectors can be expected. The SVM's classification time takes an order of magnitude longer than for our models. In addition, our model provides insight into the classification behavior due to the resulting relevance profiles. Figure 2 depicts the relevance terms λ within a local window which are achieved when training with the weighted Euclidean metric, and Figure 3 depicts the relevance terms which weight the local windows if the LIK similarity is used. The mean of the relevance terms for the 4 nucleotides at one position is depicted. Clearly, the region around a potential splice site (the possible consensus string) is of particular importance. The characteristic dinucleotide GT or AG, respectively is relevant, since we deal with the three class problem to differentiate donors and acceptors from decoys. Note that the relevance profile of both, the weighted Euclidean similarity and LIK weight the left site a bit more than the right one which might indicate the relevance of the pyrimidine rich region. In addition, the profile for LIK looks smoother and it also increases the relevance of the borders. This effect is due to the specific choice of the similarity: information is contained in more than one window since nucleotides are taken into account in different adjacent windows. The relevance terms can be distributed more smoothly among adjacent windows. The nucleotides at the borders are contained in less windows than nucleotides at the centers. As a consequence, the

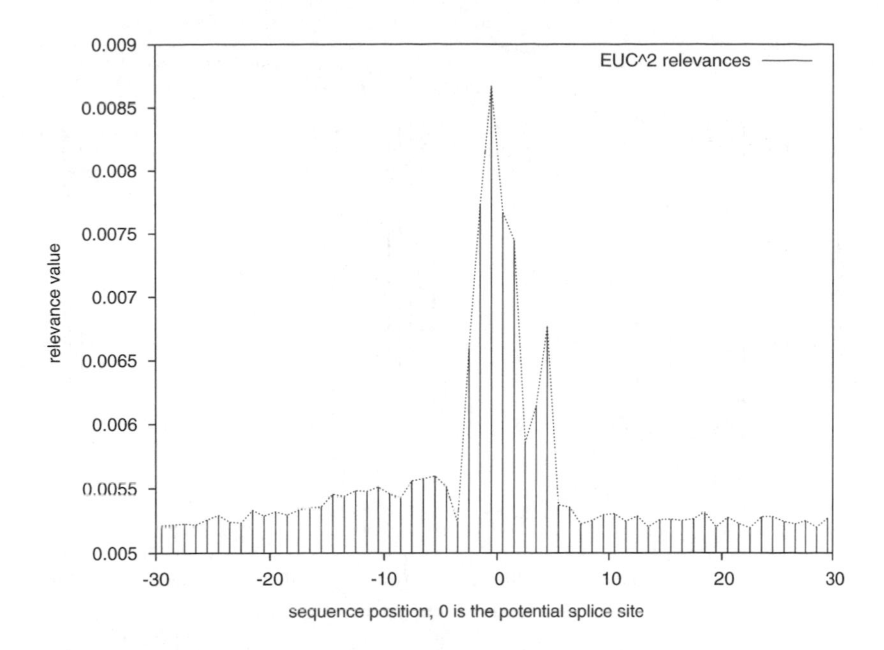

Fig. 2. Relevance profile for the weighted Euclidean similarity measure for the IPSplice data set. Each position shows the average over the the successive relevance factors for the four nucleotides at one position of the window.

relevance of local windows at the borders is to be emphasized such that nucleotides at the borders have at least the same relevance as nucleotides at the center. Note that the increase of relevance at the borders is strictly linear and it can be observed for the leftmost three windows or rightmost three windows, respectively, which directly corresponds to the smaller number of occurrences of the border corresponding to the 3D-nucleotides-embedding coordinates for LIK.

Another measure which is particularly interesting if splice sensors are to be integrated into gene finding systems is the specificity and sensitivity of the model. Specificity refers to the percentage of non-splice sites which have been predicted correctly compared to the number of decoys in the training set. Sensitivity refers to the percentage of correctly predicted splice sites compared to all potential splice sites in the data set. High specificity means that the gene finding system can assume that most decoys are correctly identified and the false-negative rate is small. High sensitivity means that almost all potential splice sites are reported, however, possibly many non splice sites are also reported such that intensive backtracking might be necessary. Obviously, these two measures constitute contradictory demands and a good balance between both is to be found depending on the properties of the whole gene finding system. For many splice site sensors, it is very easy to explicitly control the balance of specificity and sensitivity. One can vary the classification boundary and report specificity and sensitivity for several choices of the boundary. In the case

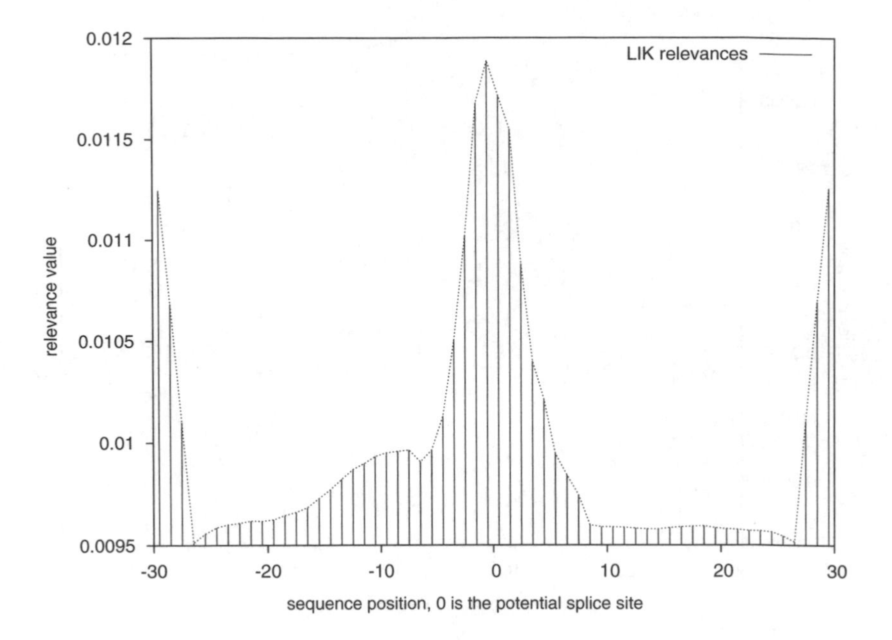

Fig. 3. Relevance profile for the locality improved similarity measure for the IPSplice data set.

of SRNG, we restrict to the two-class problem of separating splice sites (acceptor or donor) from non-splice sites. Given a data point, we can then interprete the term

$$\frac{d^\lambda_{r+} - d^\lambda_{r-}}{d^\lambda_{r+} + d^\lambda_{r-}}$$

as a measure for the likelihood of being a splice site, whereby d^λ_{r+} denotes the similarity to the closest prototype which class is a splice site, and d^λ_{r-} denotes the similarity to the closest prototype with class decoy. Note that this term lies in the interval $[-1, 1]$. Standard SRNG maps all data points with activation at most 0 to the class of splice sites. Alternative values of specificity and sensitivity can be achieved by varying the classification boundary within the interval $[-1, 1]$. A complete receiver operating characteristic (ROC) curve which depicts the false negative rate ($1-$specificity) versus the true positive rate (sensitivity) is given in Fig. 4. It is possible to achieve specificity and sensitivity 0.975 for the test set with LIK and specificity and sensitivity 0.96 for the test set and the weighted Euclidean similarity. Note that only a subset of the whole range is depicted in Fig. 4 to make the differences between the single graphs visible. A global picture including the diagonal can be found in Fig. 5.

C.elegans

For this data set the way in which decoys have been chosen is precisely documented. They are collected close to true sites. It can be expected that precise prediction of

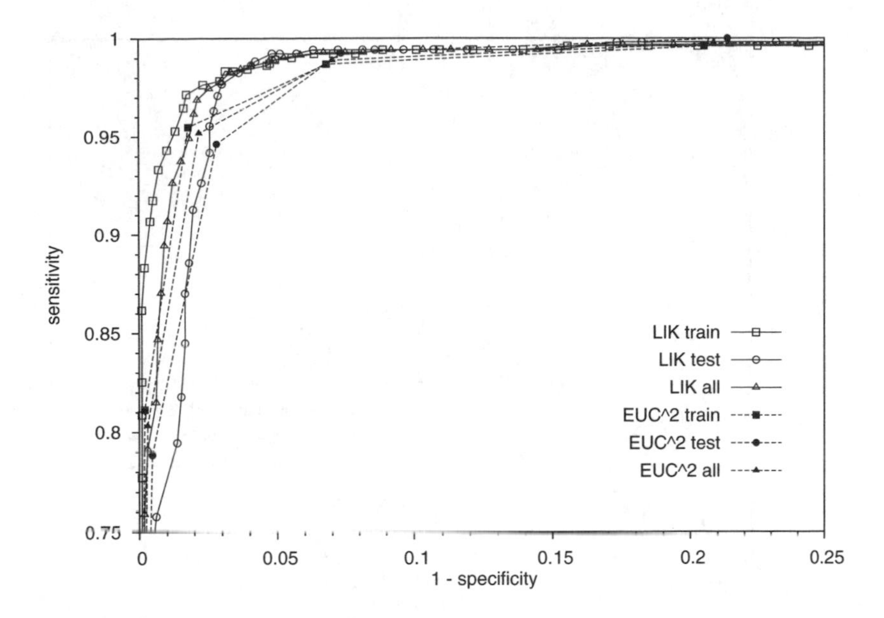

Fig. 4. ROC curve for the IPSplice data set for the two similarity measures (EUC$\wedge$2 = weighted Euclidiqan measure, LIK = locality improved similarity) and training and test set and both. Note that the graph is quite close to axes such that high specificity and sensitivity can be achieved simultaneously.

splice sites is particularly relevant (and possibly also particular difficult) for regions around potential splice sites since fast regions of intergenic DNA can be filtered out before splice sensors are applied. We trained for the same 5 data sets as the models in [63] such that we can directly compare our results with recent results for SVMs and HMMs. We used the weighted quartic metric with only two prototypes per class and parameters as follows: the learning rate for prototypes is chosen as $\epsilon^+ = 0.005$ and ϵ^- is increased from a small value to $0.05 \cdot \epsilon^+$. The learning rate for the relevance terms is $3 \cdot 10^{-8}$. Initial neighborhood range is 1 which is multiplicatively decreased by 0.9 in each epoch. We trained 600 epochs for the data set with 1000 points and 60 epochs for the data set with 10000 points. The achieved classification accuracy for the test set is $95.2\% \pm 0.176\%$ for the result of the first training set and $95.688\% \pm 0.09\%$ for the second one. In addition to these results we used a procedure which similarly to LIK takes local correlations into account: data are encoded explicitly as adjacent dinucleotide pairs which are at most one nucleotide apart. I.e. given a point x, for each position i of the window around a potential splice site the pairs (x_{i-2}, x_i), (x_{i-1}, x_i), (x_i, x_{i+1}), and (x_i, x_{i+2}) are considered. Each pair has 16 potential choices, which are encoded in a unary way for each pair. We thus obtain input vectors of dimensionality 1520 where local correlations within a neighborhood of radius 2 of each nucleotide are explicitly encoded. For this more complex encoding (which we also refer to as LIK since it is very similar in spirit to LIK but allows

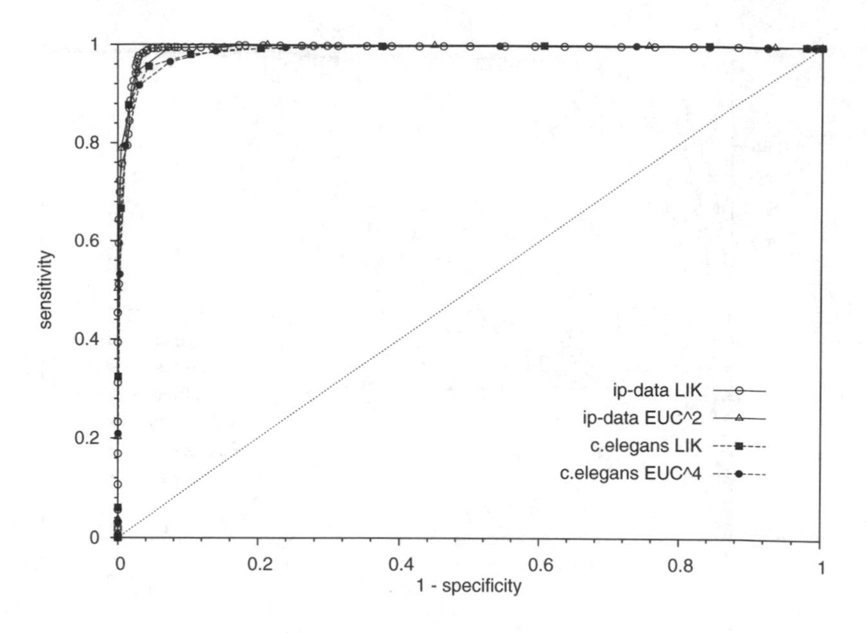

Fig. 5. ROC curve for the IPSplice and c.elegans data sets (test sets) for the respectve different similarity measures in comparison to the diagonal.

more flexibility due to separate weighting of all entries), we use the quartic weighted metric and SRNG. Here we choose 5 prototypes per class and the following parameters: learning rate $\epsilon^+ = 0.015$ for correct prototypes and $\epsilon^- = 0.01 \cdot \epsilon^+$ for the closest incorrect prototype. $3 \cdot 10^{-8}$ for the relevance terms, and initial neighborhood range 4 which is not decreased during training. Training has been done for 600 epochs for the smaller sets and 100 epochs for the larger sets. The achieved accuracy on the test sets is $95.2\% \pm 0.176\%$ for the sets with 1000 training examples and $95.688\% \pm 0.09\%$ for the sets with 10000 training examples. Results for SVM and HMM can be found in Table 2.

As indicated in Table 2, our results are competitive to the results achieved with the SVM and the locality improved kernel. Results for statistical models and combinations with SVM, however, show a better performance in particular for the large data set. The variance of these methods is a bit higher. This can be explained by the fact that for large enough data sets the statistical models can capture more relevant information from the training data such that problem adapted kernels arise. We could, of course, use these statistical similarity measure also for our approach and would likely achieve improved results as well. However, the training time for statistical models is usually quite demanding. Training one SRNG model in our experiments took only about one hour on a Pentium III (700 MHz) computer for 10000 data points whereas training a statistical model and SVM might take a day or more. Even more interesting is the complexity of classification. Our models are rather sparse since they are given by only 10 prototypes and classification of the whole test set of size 10000

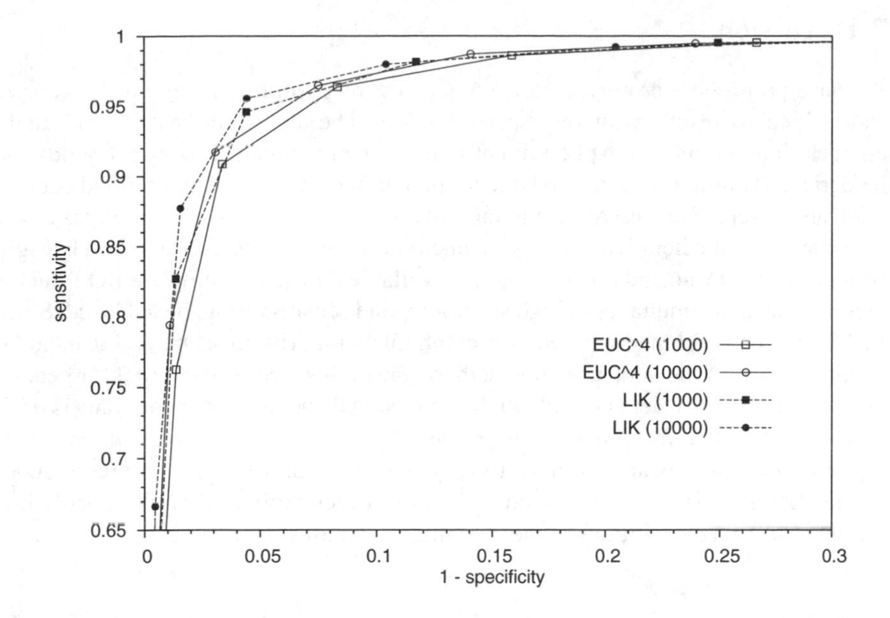

Fig. 6. ROC curve for the c.elegans data set (test sets) for the respective different similarity measures (EUC$\wedge$2 = weighted quartic similarity, LIK = dinucleotide encoding and weighted quartic similarity).

takes only a few seconds. For SVM, the classification time is one or two orders of magnitudes slower because of the number of support vectors which scales with the training set size. Hence our model is much more appropriate for tasks such as fast online prediction. This property would also transfer if we combined SRNG with alternative similarities derived from statistical models. If these measures were used in combination with prototype based classifiers instead of SVM much sparser models could be achieved.

We also include a complete ROC curve for our models. Note that Fig. 6 does only contain a subarea of the whole graph as depicted in Fig. 5 such that the differences of the models are better visible. We can obtain simultaneous specificity and sensitivity of about 0.95 if dinucleotide encoding is used.

method	$\text{our}_{\text{quartic}}$	our_{LIK}	HMM	SVM_{LIK}	SVM_{TOP}	SVM_{FK}
1000	94.6±0.003	95.2±0.15	97.2± 0.1	94.8±0.1	95.4±0.4	96.5±0.2
10000	94.8±0.003	95.7±0.09	97.4± 0.2	96.1±0.2	97.7± 0.1	97.5±0.3

Table 2. Accuracy (in %) of various methods achieved on the C.elegans dataset, only the test set accuracy is reported.

7 Discussion

We have proposed a prototype based splice size recognition model which achieves state of the art results with very sparse models. The algorithm has a sound mathematical foundation and explicit dimension independent generalization bounds can be derived. Training is robust and due to the incorporation of neighborhood cooperation also insensitive with respect to initialization. Note that further training is easily possible due to the iterative training scheme if new data becomes available. The algorithm has been evaluated on two publicly available data sets. Complete ROC curves demonstrate that simultaneous high specificity and sensitivity can be achieved. Since SRNG can be combined with any differentiable similarity measure and it includes adaptive terms of the similarity, the method constitutes a convenient tool for general classification tasks where possibly high-dimensional and heteroneneous data is dealt with. The classifier also provides insight into the classification because of the resulting relevance profiles and explicit (usually only few) prototypes as representatives of the classes. This fact can be used to further extract explicit (though possibly less accurate) logical descriptions of the classifier as demonstrated in [22].

References

1. Genome sequence of the Nematode Caenorhabditis elegans. *Science* 282:2012–2018, 1998. See also http://genome.wustl.edu/gsc/C_elegans for the specific chromosome and GFF files used here.
2. M. Arita, K. Tsuda, and K. Asai. Modeling splicing sites with pairwise correlations. *Bioinformatics* 18(Supplement 2):S27-S34, 2002.
3. A. Bairoch. Prosite: a dictionary of sites and patterns in proteins. *Nucleid acid reserach* 20:2013–2018, 1992.
4. P.L. Bartlett and S. Mendelson. Rademacher and Gaussian complexities: risk bounds and structural results. *Journal of Machine Learning and Research* 3:463-482, 2002.
5. A. M. Bianucci, A. Micheli, A. Sperduti, and A. Starita. Application of Cascade Correlation Networks for Structures to Chemistry. *Applied Intelligence Journal*, 12(1/2): 117–146, 2000.
6. C.L. Blake and C. J. Merz, UCI Repository of machine learning databases , Irvine, CA: University of California, Department of Information and Computer Science.
7. T. Bojer, B. Hammer, and C. Koers. Monitoring technical systems with prototype based clustering. In: M. Verleysen (ed.), *European Symposium on Artificial Neural Networks'2003*, D-side publications, pages 433–439, 2003.
8. S. Boué, M. Vingron, E. Kritventseva, and I. Koch. Theoretical analysis of alternative splice forms using computational methods. *Bioinformatics* 18(Supplement 2):S65–S73, 2002.
9. O. Bousquet and D.J.L. Herrmann. On the complexity of learning the kernel matrix. In: *NIPS 2002*.
10. V. Brendel and J. Kleffe. Prediction of locally optimal splice sites in plant pre-mRNA with applications to gene identification in Arabidopsis thaliana genomic DNA. *Nucleid Acid Research* 26:4748–4757, 1998.
11. S. Brunak, J. Engelbrecht, and S. Knudsen. Prediction of human mRNA donor and acceptor sites from the DNA sequence. *Journal of Molecular Biology*, 220:49–65, 1991.

12. C. Burge, and S. Karlin. Prediction of complete gene structures in human genomic DNA. *Journal of Molecular Biology*, 268:78–94, 1997.
13. C. Burges. A tutorial on support vector machines for pattern recognition. *Knowledge Discovery and Data Mining*, 2(2), 1998.
14. D. Cai, A. Delcher, B. Kao, and S. Kasif. Modeling splice sites with Bayes networks. *Bioinformatics*, 16(2):152–158, 2000.
15. J.-M. Claverie. Some useful statistical properties of position-weight matrices. *Computers and Chemsitry* 18(3):287–294, 1994.
16. V. Cherkassky and Y. Ma, Selecting the loss function for robust linear regression. Submitted to: *Neural Computation*.
17. Consortium. Initial sequencing and analysis of the human genome. *Nature* 409:860–924, 2001.
18. K. Crammer, R. Gilad-Bachrach, A. Navot, and A. Tishby. Margin analysis of the LVQ algorithm. In: *NIPS 2002*.
19. S. Degroeve, B. de Baets, Y. Van de Peer, and P. Rouzé. Feature subset selection for splice site prediction. *Bioinformatics* Supplement 2: S75–S83, 2002.
20. C.H.Q. Ding. Unsupervised feature selection via two-way ordering in gene expression analysis. *Bioinformatics* 19(10):1259–1266, 2003.
21. B. Hammer and K. Gersmann. A note on the universal approximation capability of support vector machines. Neural Processing Letters 17: 43-53, 2003.
22. B. Hammer, A. Rechtien, M. Strickert, and T. Villmann. Rule extraction from self-organizing maps. In: J.R. Dorronsoro (ed.), *Artificial Neural Networks - ICANN 2002*, pages 370–375, Springer, 2002.
23. B. Hammer, M. Strickert, and T. Villmann. On the generalization ability of GRLVQ-networks. Submitted.
24. B. Hammer, M. Strickert, and T. Villmann. Supervised neural gas with general similarity. Accepted at *Neural Processing Letters*.
25. B. Hammer, M. Strickert, and T. Villmann. Learning vector quantization for multimodal data. In: J.R. Dorronsoro (Ed.), *Artificial Neural Networks – ICANN 2002*, Springer, pages 370–375, 2002.
26. B. Hammer and T. Villmann. Generalized relevance learning vector quantization. *Neural Networks* 15:1059–1068, 2002.
27. D. Harn, D.D. Lee, S. Mika, and B. Schölkopf. A kernel view of the dimensionality reduction of manifolds. Submited to *NIPS 2003*.
28. D. Haussler. *Convolutional kernels for dicrete structures*. Technical Report UCSC-CRL-99-10, Computer Science Department, University of California at Santa Cruz, 1999.
29. S. Hebner, M. Alekseyev, S.-H. Sze, H. Tang, and P.A. Pevzner. Splicing graphs and EST assemply problem. *Bioinformatics* 18(Supplement 1):S181–@188, 2002.
30. S. M. Hebsgaard, P. G. Korning, N. Tolstrup, J. Engelbrecht, P. Rouze, and S. Brunak. Splice site prediction in Arabidopsis thaliana DNA by combining local and global sequences information. *Nucleid Acid Research* 24:3439–3452, 1996.
31. J. Henderson, S. Salzberg, and K. Fasman. Finding genes in human DNA with a hidden Markov model. *Journal of Computational Biology*, 4:127–141, 1997.
32. K. Hornik, M. Stinchcombe, and H. White. Multilayer feedforward networks are universal approximators. *Neural Networks*, pages 359–366, 1989.
33. T. Jaakkola, M. Diekhans, and D. Haussler. A discriminative framework for detecting remote protein homologies. *Journal of Computational Biology*, 7:95–114, 2000.
34. W.J. Kent and A.M. Zahler. The intronerator: exploring introns and alternative splicing in Caenorhabditis elegans. *Nucleid Acid Research* 28(1):91–93, 2000.

35. S. Kim, E.R. Dougherty, J. Barrera, Y. Cheng, M.L. Bittner, and J.M. Trent. Strong feature sets from small samples. *Journal of Computational Biology* 9(1):127–146, 2002.
36. J. Kleffe, K. Herrmann, W. Vahrson, B. Wittig, and V. Brendel. Logitlinear models for the prediction of splice sites in plant pre-mRNA sequences. *Nucleid Acid Research* 24(23):4709–4718, 1996.
37. T. Kohonen. Learning vector quantization. In M. Arbib, editor, *The Handbook of Brain Theory and Neural Networks*, pages 537–540. MIT Press, 1995.
38. T. Kohonen. *Self-Organizing Maps*. Springer, 1997.
39. A. Krogh. Gene finding: putting the parts together. In: M.J. Bishop (Ed.), *Guide to human genome computing*, pp.261-274, 2nd edition, Academic Press, 1998.
40. B. Lewin. *Genes VII*. Oxford University Press, New York, 2000.
41. Y. Li, C. Campbell, and M. Tipping. Bayesian automatic relevance determination algorithms for classifying gene expression data. *Bioinformatics* 18(10):1332–1339, 2002.
42. N. Mache and P. Levi. GENIO – a non-redundant eukariotic gene database of annotated sites and sequences. *RECOMB'98 Poster*, New York, 1998.
43. T. Martinetz, S. Berkovich, and K. Schulten. 'Neural-gas' network for vector quantization and its application to time-series prediction. *IEEE Transactions on Neural Networks* 4(4):558-569, 1993.
44. T. Martinetz and K. Schulten. Topology representing networks. *Neural Networks*, 7(3):507–522, 1993.
45. I. M. Meyer and R. Durbin. Comparative ab initio prediction of gene structures using pair HMMs. *Bioinformatics* 18(10):1309–1318, 2002.
46. D. Michie, D.J. Spiegelhalter, C.C. Taylor (eds). Machine Learning, Neural and Statistical Classification. Ellis Horwood 1994. http://www.amsta.leeds.ac.uk/c̃harles/statlog/
47. J. S. Pedersen and J. Hein. Gene finding with a hidden Markov model of genome structure and evolution. *Bioinformatics* 19(2):219–227, 2003.
48. D. J. Patterson, K. Yasuhara, and W. L. Ruzzo. Pre-mRNA secondary structure prediction aids splice site prediction. In: Altman et al. (eds.), *Pacific Symposium on Biocomputing*, pages 223–234, World Scientific, 2002.
49. M. Pertea, X. Lin, and S. L. Salzberg. GeneSplicer: a new computational method for splice site prediction. *Nucleid Acids Research*, 29(5):1185–1190, 2001.
50. S. Rampone. Recognition of splice junctions on DNA sequences by BRAIN learning algorithm. *Bioinformatics*, 14(8):676–684, 1998.
51. M. G. Reese, F. H. Eeckman, D. Kulp, and D. Haussler. Improved splice site detection in Genie. *Journal of Computational Biology*, 4(3):311–324, 1997.
52. B. Ripley. *Pattern Recognition and Neural Networks*. Cambridge University Press, 1996.
53. B. Rost and S. O'Donoghue. Sisyphus and protein structure prediction. *Bioinformatics*, 13:345–356, 1997.
54. S. L. Salzberg, A. L. Delcher, K. Fasman, and J. Henderson. A decision tree system for finding genes in DNA. *Journal of Computational Biology*, 5:667–680, 1998.
55. S. L. Salzberg, A. L. Delcher, S. Kasif, and O. White. Microbial gene identification using interpolated Markov models. *Nucleid Acids Research*, 26:544–548, 1998.
56. A. S. Sato and K. Yamada. Generalized learning vector quantization. In G. Tesauro, D. Touretzky, and T. Leen, editors, *Advances in Neural Information Processing Systems*, volume 7, pages 423–429. MIT Press, 1995.
57. A.S. Sato and K. Yamada. An analysis of convergence in generalized LVQ. In L. Niklasson, M. Bodén, and T. Ziemke (eds.) *ICANN'98*, pages 172-176, Springer, 1998.
58. B. Schölkopf. *The kernel trick for distances*. Technical Report MSR-TR-2000-51. Microsoft Research, Redmond, WA, 2000.

59. S. Seo and K. Obermayer. Soft learning vector quantization. *Neural Computation* 15: 1589-1604, 2003.
60. R. Staden. Finding protein coding regions in genomic sequences. *Methods in Enzymology* 183:163–180, 1990.
61. V. V. Solovyev, A. A. Salamov, and C. B. Lawrence. Predicting internal exons by oligonucleotide composition and discriminant analysis of spliceable open reading frames. *Nucleid Acids Research*, 22:5156–5163, 1994.
62. S. Sonnenburg, *New methods for splice site recognition.* Diploma Thesis, Humboldt-Universität Berlin, Institut für Informatik, 2002.
63. S. Sonnenburg, G.Rätsch, A. Jagota, and K.-R. Müller. New methods for splice site recognition. In: J. R. Dorronsoro (ed.), *ICANN'2002*, pages 329–336, Springer, 2002. Data sets at: http://mlg.anu.edu.au/∼raetsch/splice/
64. J. P. Staley and C. Guthrie. Mechanical devices of the spliceosome: motor, clocks, springs, and things. *Cell*, 92:315–326, 1998.
65. M. Strickert, T. Bojer, and B. Hammer. Generalized relevance LVQ for time series. In: G.Dorffner, H.Bischof, K.Hornik (eds.), *Artificial Neural Networks - ICANN'2001*, Springer, pages 677–683, 2001.
66. G. G. Towell and J. W. Shavlik. Knowledge-Based Artificial Neural Networks. *Artificial Intelligence*, 70(1-2):119–165, 1994.
67. V. Vapnik and A. Chervonenkis. On the uniform convergence of relative frequencies of events to their probabilities. *Theory of Probability and its Applications* 16(2):264-280, 1971.
68. T. Villmann. Topology preservation in self-organizing maps. In E. Oja and S. Kaski, *Kohonen Maps*, pages 279–292, Elsevier, 1999.
69. T. Villmann, E. Merenyi, and B. Hammer. Neural maps in remote sensing image analysis. *Neural Networks*, 16(3-4):389–403, 2003.
70. J. Weston, F. Pérez-Cruz, O. Bousquet, O. Chappelle, A. Elisseeff, B. Schölkopf. Feature selection and transduction for prediction of molecular bioactivity for drug design. *Bioinformatics* 19(6):764–771, 2003.
71. M. Zhang and T. Marr. A weighted array method for splicing and signal analysis. *CABIOS* 9:499–509, 1993.

Content Based Image Compression in Biomedical High-Throughput Screening Using Artificial Neural Networks

Udo Seiffert

Leibniz-Institute of Plant Genetics and Crop Plant Research Gatersleben, Germany
seiffert@ipk-gatersleben.de

Summary. Biomedical High-Throughput Screening (HTS) requires specific properties of image compression. Particularly especially when archiving a huge number of images of one specific experiment the time factor is often rather secondary, and other features like lossless compression and a high compression ratio are much more important. Due to the similarity of all images within one experiment series, a content based compression seems to be especially applicable. Biologically inspired techniques, particularly Artificial Neural Networks (ANN) are an interesting and innovative tool for adaptive intelligent image compression, although a couple of promising non-neural alternatives, such as CALIC or JPEG2000 have become available.

1 Introduction

Image acquisition and processing techniques [3, 12] have evolved into powerful and high-capacity tools. Particularly when linked with intelligent analysis methods, remarkably complex systems can be built. The scope of applications ranges from rather universal tasks to very specific ones and from microscopic to macroscopic level.

The importance and spread of *High-Throughput Screening* (HTS) [13] methods, being applied in almost all scientific fields, from life science to engineering, is increasingly high. All of these have in common that thousands, millions or even more, equal or similar experiments or investigations are performed. Besides the actual result of a particular experiment, which is often rather low-dimensional – sometimes even a simple yes/no decision – HTS often leads to an immense quantity of data. This may be more or less temporary, but has to be stored, for example, to be processed in a batch process or for further reference. Depending on the particular application this data is more or less high-dimensional.

Especially when dealing with image based data processing, the amount of data to be handled often becomes very large. A typical high-resolution digital image from a microscope is between 15 and 50 MBytes, i.e. 3.000 × 2.000 pixels and 16 bit per color channel leads to a net size of about 35 MBytes. Sometimes images are tiled to achieve a higher local resolution. As a technical consequence to save storage capacity or transmission bandwidth, the images are usually compressed.

In contrast to on-line compression, where data and particularly images are compressed and decompressed *on the fly*, such as on the Internet, compression time plays not a very important role when storing is rather (long-term) archiving. On the other hand, in scientific applications not only is visual correctness of the images (having a pretty picture for the human eye) required, but also any kind of distortion is usually not tolerable. Details must not be lost, since further processing is performed *after* the images have been compressed/decompressed, maybe even several times. This motivates the utilization of lossless coding schemes. However, as to be seen afterwards in Sect. 2, insisting on lossless coding limits the achievable compression ratio. In this case lossy compression methods, which retain particularly important features of the images, have to be considered. Since an analytical resp. formally mathematical definition of both image features in general, and those not to be distorted by a compression in particular, is rather difficult, adaptively trainable systems seem to be a promising approach.

Another important fact should be kept in mind when talking about HTS properties. The image content in all images of a particular HTS application is often very similar. In other words, the images to be compressed are characterized by limited variability, because they contain for instance the same type of cell formation. In this example the meaningful information is not any cell itself, but maybe slight differences of its shape or texture. This can only be recognized correctly if the images – or at least those image features required for this – are not significantly interfered with by compression artifacts. In addition to the above mentioned motivation, this opens a realistic chance to apply an image content based compression. Besides some fractal approaches [10], Artificial Neural Networks (ANN) have been used very successfully [7, 5, 29, 1].

This chapter reviews standard as well as intelligent and adaptive image compression algorithms in the context of High Throughput Screening. It demonstrates the features of different approaches by means of a number of real-world images from biomedical investigations.

2 Reviewing Standard Image Compression Algorithms

As a matter of fact, almost all standard image compression algorithms are linked to at least one particular image file format. Some file formats provide just one compression method, while others can be considered as some kind of a container – just building a wrapper around a number of different applicable compression methods. As already mentioned in Sect. 1, a lossless or nearly lossless compression is desired to meet the requirements of this kind of scientific image processing. In this context it seems to be unnecessary to point out that a color depth of at least 3×8 bit is essential. This section reviews the supply of file based standard image compression algorithms.

Having set up these constraints, there are several standard image file formats (see Table 1) available, offering a more or less wide variety of internal compression algorithms [18, 16]. While RLE and LZW based compression [30] is originally lossless, JPEG [20] is a lossy compression format. It still offers a reasonable and finely adjustable balance of retained quality and gained compression. The newer JPEG2000 [27] can be both lossy and lossless (nearly lossless) and outperforms JPEG and many other methods.

Table 1. Selection of standard file formats and corresponding compression schemes as well as their suitability for storing image data in a High Throughput Screening environment. For comparison one non file based compression algorithm (CALIC) is mentioned.

File format	Compression algorithm	Max. color depth	HTS suitability
Bitmap (*bmp*)	None	3 × 8 bit	–
	Run Length Encoder (*RLE*)	3 × 8 bit	+
Graphics Interchange Format (*gif*)	Lempel-Ziv Welch (*LZW*)	1 × 8 bit	–…– –
Joint Pictures Expert Group (*jpg*) or (*jpeg*)	Discrete Cosine Transformation (*DCT*)	3 × 8 bit	–
JPEG2000 (*jp2*)	Discrete Wavelet Transformation	3 × 8 bit	+
		3 × 16 bit	++
Portable Network Graphics (*png*)	ZLib, CRC-32	3 × 16 bit	++
Tagged Image File Format (*tif*) or (*tiff*)	None	3 × 16 bit	–
	Lempel-Ziv Welch (*LZW*)	3 × 16 bit	++
–	Content-based, Adaptive, Lossless Image Coding (*CALIC*)	3 × 16 bit	+

All of these image file formats have more or less different properties. When evaluated in a particular application context, some properties are rather advantageous and some tend to be disadvantageous. In general, all image file formats have their strengthes when utilized in the right way. In HTS however, *png* and *tif* are most suitable due to their lossless compression feature. The *tif* format is more frequently implemented in commercial image acquisition software.

Unfortunately a number of compression algorithms which are scientifically interesting, perform very well and are even versatilely applicable, could not succeed as a (commercial) file format. Thus, in addition to the above mentioned standard methods, there are a number of non file based algorithms, like for example EZW (Embedded image coding using Zerotrees of Wavelet coefficients) [26], SPIHT (Set Partitioning in Hierarchical Trees) [22] and LMSE (Least Mean Square Encoding) [6] or LOCO-I (LOw complexity, COntext-based) [31] and CALIC (Content-based, Adaptive, Lossless Image Coding) [32], which even adapt themselves to the image using context-based predictors. Since these algorithms are often not available in standard image processing software, they are not used very frequently. However, because CALIC is considered by the image compression community as the best approach for lossless content based coding, it is mentioned for comparison in the following sections. See also Table 1.

Depending on the image acquisition hardware actually used, mainly the CCD or CMOS chip [11, 19] of a digital camera, there are also some proprietary *raw* formats available. Since the image acquisition system simply stores the raw data that comes directly off the chip, it is generally system (vendor) dependent. A special software is required to transform the image to more versatile formats, i.e. *tif*, or to prepare it for further processing. Although this is the most basic and by the way also a technically very interesting way to store a digital image (not even the interpolation of the four monochrome photo elements making up one color pixel is performed), it is not suitable for general purpose HTS applications.

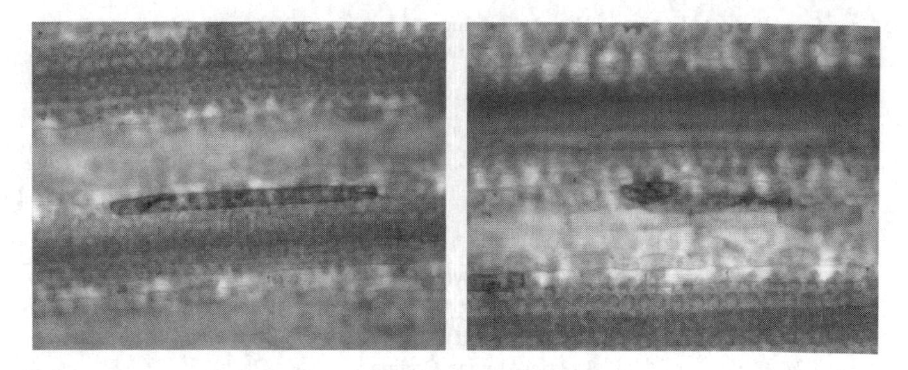

Fig. 1. Two sample images of the test data set showing microscope images of barley leafs. The images were taken using a *Zeiss AxioVision* system. The original image size is 2600×2060 pixel with a color depth of 3×8 bit.

2.1 Standard Lossless Image Compression in Practical Operation

This subsection summarizes some measured values of standard file based lossless compression algorithms in terms of the compression ratio. For this investigation and for those in the following sections as well, a set of biological and medical images was used. Two samples of the extensive data set are shown in Fig. 1.

The results shown in Table 2 indicate that *png* and *tif* (16k LZW) outperform all other file based lossless variants. Both are able to compress the images by more than 50 percent. Even better compression rates are achieved with CALIC. Since the algorithm can only process one color plane at once, the original images have to be split into separate portions. Commonly these are the RGB (red, green, blue) components, although other color spaces [3] are possible, too. While RGB as the native representation of digital color images is an additive color representation, the subtractive CMYK (cyan, magenta, yellow, black) decomposition, which is widely used in printing machines, is shown for comparison. The comparatively short computation time necessary to split the images into components has been ignored so far. A more sophisticated and possibly image content adaptive splitting of the original images into virtual color planes, based for example on a Karhunen-Loeve Transformation (KLT), leads to slightly better compression rates, but at the cost of a disproportionately long additional computation time. The results obtained with other images of the test data set are very similar.

Table 2. Lossless image compression: file size and resulting compression ratio based on the test image in Fig. 1 (left). The given file sizes may vary slightly depending on the software implementation used.

Image file format	Resulting file size	File size ratio	Comment
net	16.068.000 bytes	1.00	net image size, without file header
tif	16.084.696 bytes	1.00	original microscope file
bmp	16.068.054 bytes	1.00	without compression
bmp	11.472.828 bytes	0.71	RLE compression
png	6.367.697 bytes	0.40	
tif	15.452.899 bytes	0.96	LZW compression (8k block) [18]
tif	7.319.552 bytes	0.46	LZW compression (16k block) [18]
–	6.201.032 bytes	0.39	CALIC compression (CMYK)[1]
–	5.809.865 bytes	0.36	CALIC compression (RGB)[2]

2.2 Standard Lossy Image Compression

Lossy compression (i.e. JPEG) can squeeze the images to nearly any desired size at the cost of more or less distinct image degradation. File based standard lossy image compression usually leads to a reconstruction error, which is not tolerable for subsequent image analysis.

However, if there was a method which led to an image degradation that is not dictated by the algorithm but could be specified anyhow by the user and was dependent on what image content is really important, even lossy compression could be

[1] The color images were decomposed into 4 separate color planes CMYK (cyan, magenta, yellow, black) and then separately processed with CALIC.

[2] Ditto, but with the 3 RGB (red, green, blue) color planes.

applicable. These issues are raised in the next section and discussed in the following ones.

3 Intelligent Biologically Inspired Image Compression

Universal compression algorithms usually process all images equally. In other words, possibly existing a-priori information on the image content is not passed to the compression algorithm. In general it seems to be difficult to gather this information at all. Furthermore, this would limit the versatility – resulting in a hardly manageable number of very specialized compression methods and image file formats. For most cases existing file based compression methods meet the requirements very well. After all, who wants to analyze the image content before transmitting it via the internet?

A different situation can be found in image based HTS. Here all images often contain very similar information, and since there is a very high number of images to be processed, it seems to be worthwhile and sometimes even necessary to search for more sophisticated solutions.

In many successful implementations artificial neural networks have proven to provide a key feature which enjoys all the advantages of an intelligent problem solution. Now intelligent image compression seems to be another challenging application for neural nets [5].

3.1 Auto-associative Multiple-Layer Perceptrons

The most straightforward method of applying neural nets to image compression is the utilization of Multiple-Layer Perceptrons (MLP) [29].

MLPs [17, 21, 15] are probably the most frequently used supervised trained neural nets. Setting them into an *auto-associative* mode (Fig. 2), they are able to produce an output vector which is, apart from a small remaining error, identical to the currently presented input vector. In other words, the network is supposed to show the presented input vector at its output, which means nothing less than an overall transfer function of 1.

Assuming the length m of the hidden layer is smaller than the length n of both input and output layers, all information is compressed by the ratio m/n while passed from input to hidden layer. The inverse operation is performed between the hidden and output layers. In contrast to the more common hetero-associative mode of feed-forward nets, the actually meaningful result, namely the compressed input vector, is not provided by the output (as usually) but by the hidden layer.

The network may contain one or three hidden layers. In the latter case the compression/decompression is made in two steps each and the result is provided by the middle hidden layer. The additional hidden layers (1 and 3) are usually of the equal length k with $m \leq k \leq n$. In general all layers of the net are fully connected.

A feed-forward network run in auto-associative mode can now be applied for image compression too. The input image is divided into blocks of typically 8×8

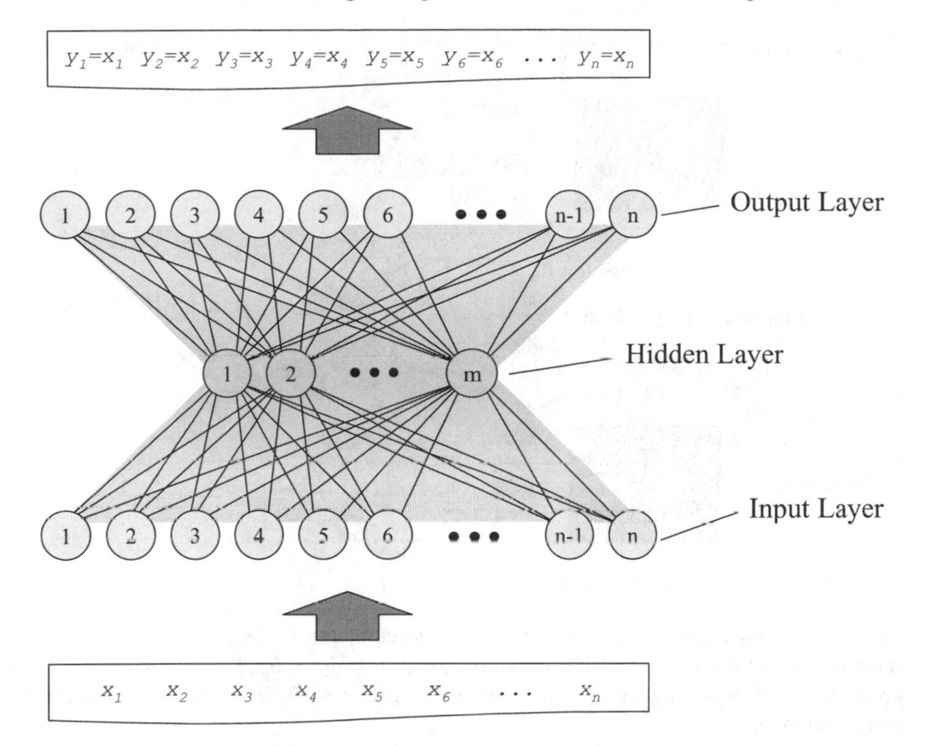

Fig. 2. Auto-associative Multiple Layer Pertceptron with n nodes within the input and output layer as well as m nodes within the hidden layer. Basically is $m \neq n$, but for compression tasks is $m < n$.

pixels (similar to JPEG) and then rearranged into a vector of length 64 (for grayscale images or just one color plane) or 192 containing all color triples (see Fig. 3). In general, all RGB color planes can be processed either by three separate networks of identical topology but different weights or by one bigger network.

In case of just one hidden layer, each weighted connection can be represented by $(w_{i,j}, i = 1, 2, \ldots, n$ and $j = 1, 2, \ldots, m)$ or by a weight matrix of $n \times m$ and $m \times n$ respectively. While the network is trained to minimize the quadratic error between input and output, the weights are internally changed to find an optimal average representation of all image blocks within the hidden layer. In connection with the nonlinear activation function f of the neurons, an optimal transformation from input to hidden layer and vice versa from hidden to output layer has to be found during the network training by changing the weights $w_{i,j}$.

For a single normalized pixel $p_i \in [0, 1]$ of the original image this transformation can be written as

$$h_j = f(\sum_{i=1}^{n} w_{i,j} p_i) \quad \text{with } (1 \leq j \leq m) \tag{1}$$

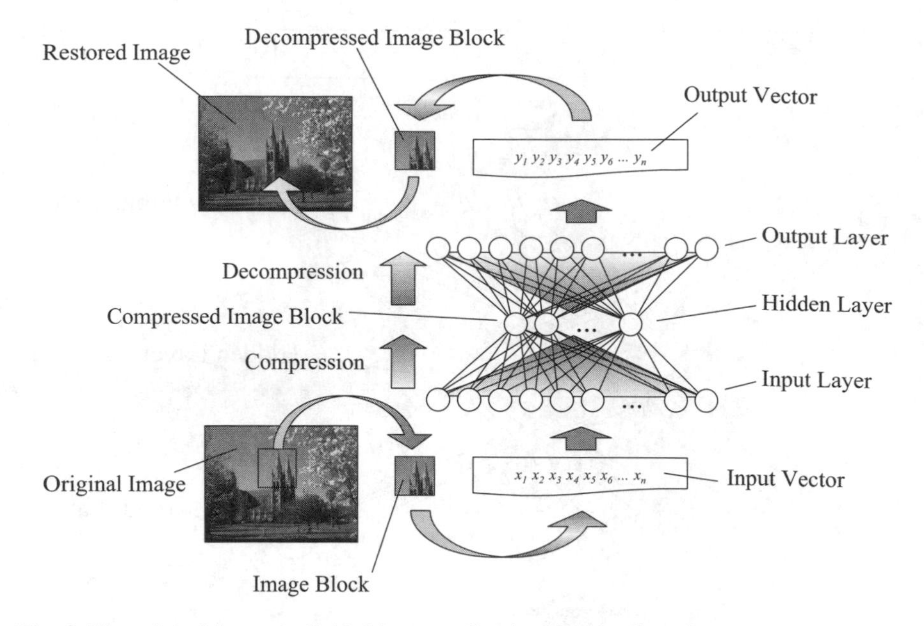

Fig. 3. The original image is divided into equally sized blocks ($b_x \times b_y$). All blocks are consecutively transformed into an input vector of length $n = b_x \times b_y$. The hidden layer is built up of $m \leq n$ neurons and contains the compressed image block. The ratio m/n denotes the compression rate.

for the compression and

$$p'_i = f(\sum_{j=1}^{m} w'_{j,i} h_j) \quad \text{with} \quad (1 \leq i \leq n) \tag{2}$$

for the decompression.

The ratio m/n denotes the compression rate. We need n memory positions to store one block of the original image but only m positions for a compressed one. Since the number of hidden nodes m can be set up arbitrarily, the compression rate can be a-priori specified within this range at all possible quantization steps $(1/n, 2/n, 3/n, \dots, (n-1)/n)$.

3.2 Extension to an Adaptive Intelligent Image Compression

The key motivation for the utilization of content based image compression as mentioned in the introductory section, namely the similarity of all images within a series of investigations, pushes even more for an adaptation of the compression algorithm itself. This goes beyond the neural networks based method described in the previous section. The idea is to analyze the content of an image or a currently processed part of it and then select a very specialized compressor.

In order to put these things together, the core statements as aforesaid are repeated:

1. File based compression methods, as described in Sect. 2, treat all images the same way, regardless of their content. This is the most universal method and exactly what is to be expected from a versatile graphics file format.
2. The neural networks solution also works without respect to different content of the processed image blocks. However, in most cases this is already more specialized, since the training data set inevitably contains images with limited variability. By selecting the training images, some kind of hand-operated specialization is performed, because the weights are set up with respect to the presented training examples. If particular structures constantly occur, the MLP will remember these very well.

Although in a completely different context, there are some investigations demonstrating that image blocks of relatively small size (up to 32×32 pixels) contain sufficiently limited information to be sorted into a small number of classes [24]. This is the major condition to successfully implement an image block adaptive compression system. The general idea is shown in Fig. 4.

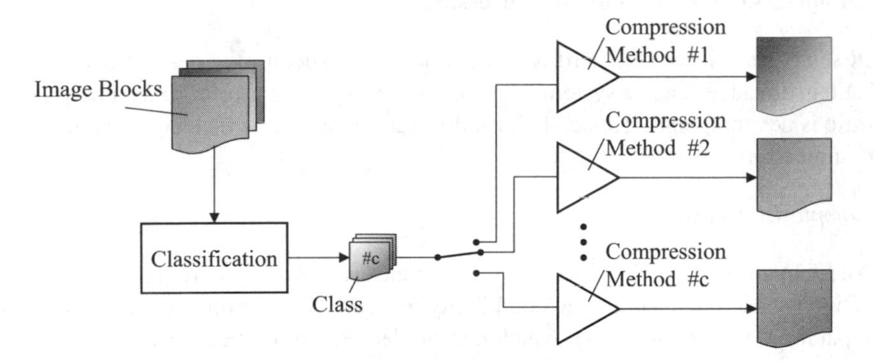

Fig. 4. Based on the image blocks (see Fig. 3) a classification by means of extracted image properties (i.e. similarities on image level) is performed. Depending on its class, each block is processed by a specialized compression method.

Each image block is processed by a specialized compression method which is optimized for the members of a particular class. There is a strong correspondence between a set of blocks belonging to the same class and the applied compression method. As long as the number of classes is low enough, the system seems to be feasible. There is, as often, a trade-off – here between good manageability (few classes) and high quality (many classes). If, as one extreme, there was only one class, the system is virtually identical to the stand-alone MLP as described in the previous section. Having as many classes as different image blocks denoted the other extreme, which is definitely not manageable. The more similar the single images[3] within one particular image data set, the better this system works – at the same number of classes a better result can be expected or, vice versa, at the same quality level less classes are necessary.

In general, for the task of the image compression any method which can be adapted or adapts itself to limited image content can be reasonably applied. Against the background of an entirely biologically motivated system, a combination of several neural networks seems to be of particular interest. The compressors can be formed, for example, from the above mentioned auto-associative Multiple-Layer Perceptron.

The (neural) classification systems have to be trained in an unsupervised manner. From the system model in Fig. 4 it can be derived and the next sections will also show that the classification task seems to be the core of the entire adaptive image compression system. Looking at the parameters which control each particular compression method leads to a number of possible approaches for the classification.

Compression Ratio Adaptive Processing

In some cases it may be hard or not applicable to specify the desired compression ratio in advance. Then a system which does not expect an a-priori fixed compression ratio is advantageous. This calls for a different parameter which controls the image compression.

Complexity Measure

An adaptation, depending on a complexity measure, solves this problem. The classification is based on a combined complexity measure (i.e. entropy, object-background separation, texture analysis), which can be derived from image features. Although there is no standard definition, usually image complexity is defined as some ratio of background homogeneity and foreground clutter [9]. The main problem is to find image features which describe its complexity with regard to a *compression relevant complexity*[4]. Furthermore, the computational effort should be reasonable, which makes an extensive feature extraction, just to obtain a complexity measure, rather impossible.

Variable Image Decomposition

As an extension of the above described method with fixed-sized blocks, a variable image decomposition can be implemented. A complexity measure is used to find an

[3] The similarity of the single images is the most important factor in this context, although a more or less distinct homogeneity of the image blocks within one image has an influence as well.

[4] For example: the size of a *jpg* file depends on the complexity of the image, but only in connection with the Cosine transformation being used for compression in *jpg*. A different compression algorithm may require a different complexity measure.

optimal size of the image blocks. These blocks do not have to be analyzed further, because a complexity measure has already (indirectly) been applied.

Apart from finding a relevant and manageable complexity measure, it seems to be difficult to set up the compressors. Since the compression ratio of the MLP is fixed by the number of neurons in the middle hidden layer (see Sect. 3), the single compressors must have a different neural topology. In case of variable blocks even the width of input / output layers is not fixed.

Similarity Adaptive Compression

Although a variable compression ratio, as described in the previous subsection, may have some advantages, its implementation and parametrization is rather difficult. And after all, a fixed compression ratio has advantages, too:

- a more precise estimation of required storage capacity is possible,
- image blocks do not have to be classified according to a hardly manageable complexity measure but by an easily implementable similarity criterion,
- compressor resp. neural network topology is fixed.

The entire system, consisting of classification and compression / decompression, can be set up completely by a couple of neural networks of two different architectures.

Classification

The classification is based on similarities at image level. This relieves the system of complex calculations. A Self-Organizing Map (SOM) can be used to perform the classification [14, 25]. For that purpose the SOM is unsupervised trained with a number of typical images. After this training, some classes are formed, which can now be used to pass the image blocks to the corresponding compressor.

Other unsupervised neural architectures, i.e. Adaptive Resonance Theory (ART) [4, 2], seem to be suitable as well, but have not been tested yet.

The actually utilized classification system determines the properties of the allocation of classes as well as the distribution of image blocks to a corresponding class. Figure 5 shows the effects of both a proper and a deliberately false classification using an SOM.

Compression / Decompression

All compressors are realized by the auto-associative Multiple-Layer Perceptrons described in Sect. 3. Since each network receives only blocks of the same class, it develops into an *expert* for a particular and very selective image (block) content. This way, the reconstruction error can be kept rather low (see Sect. 4). In general any supervised trained auto-associative (neural) system could be used. An interesting alternative approach is Support Vector Machines [28, 8, 23].

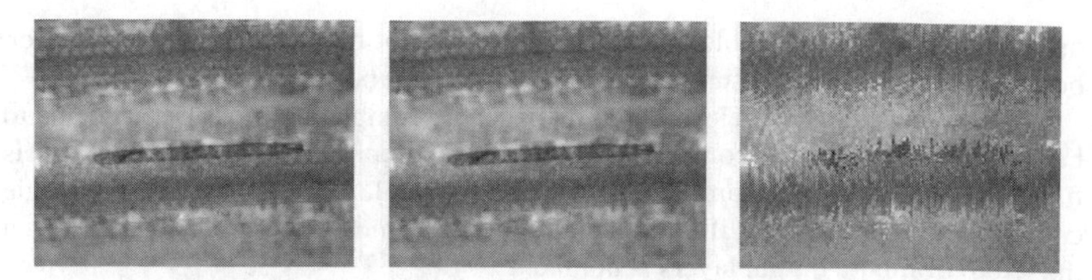

Fig. 5. Demonstration of the importance of a proper classification. From left to right: original image, restored image with correctly assigned class (optimized coder), restored image with randomly chosen class (non optimized coder).

4 Lossy Image Compression – Results

While Table 2 in Sect. 2 summarized the results for *lossless* compression, this section compares practical results of *lossy* compression for both *file based* and *neural systems*. The key question, whether the occurring image degradation caused by a lossy compression method is still tolerable at all, and if so, to what extend, has to be answered in the context of the image processing next in line (succeeding the compression / decompression). Some problem dependent fine-tuning will definitely be required.

4.1 Quality Criterion

A Root-Mean-Square (RMS) error has been used to evaluate the reconstruction quality. Because the RMS error is the de-facto standard, this seems to be suitable and sufficient.

Strictly speaking other criteria which are more relevant to the image reconstruction task, for example maximal deviation, are conceivable and maybe more qualified for particular analysis purposes. If image degradation has to be judged relating to a special application driven criterion, the standard RMS measure can be substituted by a more specialized one. Also some more biologically relevant measures motivated by human-physiological features or properties could be feasible.

The RMS measure considers all occurring image degradation the same way, regardless of its impact on further processing steps (analysis of compressed and reconstructed images). This is appropriate *exactly* and *only* for those systems which treat all image content the same way. However, though the adaptive variants lead to a difference between original and restored image which is correctly measured by the RMS, it is not equally relevant for further processing. Consequently the RMS measure overrates the occurring errors of the adaptive methods compared to the non-adaptive ones.

4.2 Experimental Set-up

The image data base is still the same as described before. For samples see Fig. 1. The results for reconstruction errors and partly also for the compression ratio depend slightly on the images actually used but indicate the general trend. However, significantly different images may lead to varying results, especially with the adaptive methods. The computation time[5] strongly depends on numerical implementation details and will plainly vary. It should only be taken as a pretty rough clue.

The *stand-alone MLP network* has been tested with 8×8 image blocks and a single hidden layer ($8 \ldots 32$ neurons) as well as three hidden layers ($\langle 16-8-16\rangle \ldots \langle 32-16-32\rangle$ neurons). However, the results (see next subsection) will show that one hidden layer is sufficient. More layers just slow down the compression / decompression with no significant improvement in reconstruction errors.

The first *compression ratio adaptive* method has been implemented with *entropy* as complexity measure. The second one is based on a *tree-structured variable image decomposition*.

The *similarity adaptive* variant uses a Self-Organizing Map (SOM) to detect the similarities of the image blocks. The SOM size has been varied between 3×3 and 7×7, the corresponding MLP networks as described above.

In competition to the neural based methods, JPEG and JPEG2000 as lossy respectively nearly lossless algorithms, have been tested with the same images. The JPEG quality coefficient was set to *maximum* and the JPEG2000 was run with the *-lossless* option. For example, the left image of Fig. 1 leads to a Frobenius norm of the color triple of $[179, 179, 516]$ at an image size of 5.356.000 pixels per color channel.

4.3 Numerical Values

As the numerical values given in Table 3 demonstrate, the best results regarding the reconstruction error are offered by JPEG2000. This seems to be obvious due to its smooth transition from arbitrarily adjustable lossy compression up to a lossless mode. Using the *-lossless* option, JPEG2000 typically leads to a compression of $1:3$. This seems to be appropriate, if not for its rather slow processing (about 10 seconds). Significantly faster (and at the same time offering a higher compression) are the tested adaptive methods. However, these are accompanied by a worse reconstruction error.

Though the stand-alone MLP is fast, it is outperformed even by the old JPEG algorithm. The differences between the several adaptive variants are not very significant. Within a certain but relatively wide range, each of the adaptive methods can be adjusted according to the user's requirements. Thus, the major differentiating factor is rather their different handling, considering the statements in Sect. 3.2.

---

[5] For the neural networks based methods it is the recall time (on-line).

[6] Since the measured values of computation times are strongly dependent on software implementations as well as hardware details, the given numbers are intended as rather a rough guide to show an order of magnitude which can be expected with common computer architecture.

Table 3. Results of different *lossy* or *nearly lossless* compression methods. In order to achieve an adequate statistical reliability, a number of similar images of the same data set was used in addition to the shown test images (see Fig. 1). The results are either simply averaged (in case of low variance) or the range is given explicitly.

Compression method	Image size ratio	Reconstruction error (RMS), approx.	Compression time [sec][6]
Stand-alone MLP	0.50	$3 \cdot 10^{-3}$	≈ 1
	0.125	$5 \cdot 10^{-3}$	< 1
Compression ratio adaptive			
(complexity measure)	$0.10 \ldots 0.25$	$8 \cdot 10^{-4} \ldots 2 \cdot 10^{-3}$	≈ 3
(variable image decomp.)	$0.10 \ldots 0.45$	$4 \cdot 10^{-4} \ldots 9 \cdot 10^{-4}$	$4 \ldots 6$
Similarity adaptive	0.125	$9 \cdot 10^{-4} \ldots 3 \cdot 10^{-3}$	≈ 1.5
JPEG (max. quality)	$0.15 \ldots 0.20$	$6 \cdot 10^{-3}$	< 1
JPEG (max. qual., progr.)	$0.17 \ldots 0.23$	$6 \cdot 10^{-3}$	< 1
JPEG2000	≈ 0.30	$5 \cdot 10^{-5}$	10

5 Discussion

Within the context of Soft Computing paradigms used for Bioinformatics this chapter presents a review of file based image compression methods as well as some intelligent neural networks based adaptive systems. All methods are judged against the background of image based biomedical High-Throughput Screening (HTS). The most characteristic properties of this field can be defined as

- images are to be processed in a huge number,
- images are often very similar,
- images are often rather homogeneous,
- images are to be archived, often without tough time constraints,
- images are not to be significantly distorted.

Since the archived images are subject to further scientific processing, usually a *noticeable* image degradation caused by coding is not acceptable. The importance but also the difficulties in this connection are located in the term *noticeable*. This fact addresses two problems – how to specify a suitable and primarily relevant measure for image degradation and then how to quantitatively determine it.

Depending on the specific task of a particular HTS application, either the reconstruction error of decoded images or the computation time for compression / decompression is the more crucial point. In every case, a trade-off between compression ratio, reconstruction error and computation time has to be found.

The most straightforward approach is to use a lossless image file format, where LZW based *tif* leads to the best compromise of compression ratio, computation time and availability. However, a compression of more than about half the original image size is not possible for typical biomedical images. If the application is not bound to a file based lossless compression, with CALIC an improved compression up to about $\frac{1}{3}$ of the original image size becomes available.

If a higher compression is desired, lossy image compression with an inevitable degradation of the original images is the only solution. The key question at this point is (as mentioned above), whether the Root-Mean-Square based image degradation measurement underrates some of the advantages of the image content adaptive methods. Unfortunately this could only be answered in the context of a particular conceptual formulation and not on a general scale.

Interestingly, but after a closer look at the details not at all surprisingly, the apparently best quality is not achieved by one of the adaptive and content based methods, but rather utilizing the wavelet based nearly lossless JPEG2000 coding. It is 1 to $1\frac{1}{2}$ decimal powers better and very easily manageable. In spite of its pretty good and probably in almost all practical applications acceptable reconstruction error, it offers compressed images of less than about $\frac{1}{3}$ of their original size. So far it seems to be the best solution. However, it is rather slow compared to all other investigated methods. That means, besides the above discussion about the degradation measurement, the more the initially mentioned trade-off turns out to be time critical, the more relevant becomes one of the suggested adaptive methods.

Apart from considering the computation time it turns out that for even higher compression rates the lossy variant of JPEG2000 or (the somewhat faster) adaptive intelligent algorithms using artificial neural networks may be the only solution. Both are very well scalable. JPEG2000 and the stand-alone as well as the similarity adaptive Multiple-Layer Perceptrons have the advantage of offering an a-priori adjustable fixed compression rate. This may be a real advantage if a particular image data set must not exceed a fixed storage capacity.

In a few words, image compression in biomedical HTS environments can be divided into three domains:

1. lossless compression (tif, png, CALIC), if the desired compression ratio does not exceed about $0.4 \dots 0.35$,
2. wavelet based compression (JPEG2000) for higher compression ratios, if image degradation is generally tolerable,
3. image content adaptive compression (neural networks) for faster lossy compression with predominantly noncritical distortions (user defined).

Acknowledgements

The author would like to thank Tobias Czauderna for supplying some software and Joseph Ronsin for a productive discussion. This work was supported by a grant of the German Federal Ministry of Education and Research (No. 0312706A).

References

1. C. Amerijckx, M. Verleysen, P. Thissen, and J.-D. Legat. Image compression by Self-Organized Kohonen Map. *IEEE Transactions on Neural Networks*, 9(3):503–507, 5 1998.
2. G. Bartfai. Hierarchical clustering with ART neural networks. In *Proceedings of the IEEE 1994 International Conference on Neural Networks*, volume 2, pages 940–944. IEEE Press, 1994.
3. G. A. Baxes. *Digital Image Processing: Principles and Applications*. John Wiley & Sons, Hoboken, NJ, 1994.
4. G. A. Carpenter and S. Grossberg. The ART of adaptive pattern recognition by a self-organizing neural network. *IEEE Computer*, 21(3):77–88, 1988.
5. S. Carrato. Neural networks for image compression. In E. Gelenbe, editor, *Neural Networks: Advances and Applications 2*, pages 177–198. Elsevier North Holland, Amsterdam, 1992.
6. Y.-S. Chung and M. Kanefsky. On 2-d recursive LMS algorithms using ARMA prediction for ADPCM encoding of images. *IEEE Transactions on Image Processing*, 1(3):416–422, 1992.
7. G. Cottrell, P. Munro, and D. Zipser. Image compression by Backpropagation: An example of extensional programming. In N. Sharkey, editor, *Models of Cognition: A Review of Cognition Science*, pages 297–311. Intellect, Norwood, NJ, 1990.
8. N. Cristianini and J. Shawe-Taylor. *An Introduction to Support Vector Machines and Other Kernel-based Learning Methods*. Cambridge University Press, Cambridge, UK, 2000.
9. G. Earnshaw. Image complexity measure. Technical Report CT92-0015, Centre for Intelligent Systems, University of Plymouth, 1995.
10. Y. Fisher. *Fractal Image Compression. Theory and Application*. Springer Verlag, Telos, 2. edition, 1996.
11. G. C. Holst. *CCD Arrays, Cameras and Displays*. Encyclopaedia Britannica, Chicago, Il, 2. edition, 1998.
12. A. K. Jain. *Fundamentals of Digital Image Processing*. Prentice Hall, Upper Saddle River, NJ, 1998.
13. W. P. Janzen. *High Throughput Screening: Methods and Protocols*. Humana Press, Totowa, NJ, 2002.
14. T. Kohonen. *Self-Organizing Maps*. Springer Verlag, London, 3. edition, 2001.
15. R. P. Lippmann. An introduction to computing with neural nets. *IEEE ASSP Magazine*, 4(87):4–23, 1987.
16. J. Miano. *Compressed Image File Formats: JPEG, PNG, GIF, XBM, BMP*. Benjamin Cummings / Addison Wesley, San Francisco, Ca, 2002.
17. M. Minsky and S. Papert. *Perceptrons: An Introduction to Computational Geometry*. MIT Press, Cambridge, 1969.
18. J. D. Murray, W. Vanryper, and D. Russell. *Encyclopedia of Graphics File Formats*. O'Reilly UK, Cambridge, 1996.
19. R. Pallas-Areny and J. G. Webster. *Sensors and Signal Conditioning*. Wiley, Hoboken, NJ, 2. edition, 2000.
20. W. B. Pennebaker and J. L. Mitchell. *JPEG: Still Image Data Compression Standard*. Kluwer International, Dordrecht, 1992.
21. D. E. Rumelhart, G. E. Hinton, and R. J. Williams. Learning internal representations by error propagation. In D. E. Rumelhart and J. L. McClelland, editors, *Parallel Distributed Processing: Explorations in the Microstructure of Cognition*, volume 1, pages 18–362. MIT Press, Cambridge, Ma, 1986.

22. A. Said and W. A. Pearlman. A new, fast and efficient image codec based on set partitioning in hierarchical trees. *IEEE Transactions on Circuits and Systems for Video Technology*, 6(3):243–250, 1996.
23. B. Schölkopf and A. J. Smola. *Learning with Kernels: Support Vector Machines, Regularization, Optimization, and Beyond.* MIT Press, Cambridge, Ma, 2001.
24. U. Seiffert. Growing multi-dimensional Self-Organizing Maps for motion detection. In U. Seiffert and L. C. Jain, editors, *Self-Organizing Neural Networks: Recent Advances and Applications*, volume 78 of *Studies in Fuzziness and Soft Computing*, pages 95–120. Springer-Verlag, Heidelberg, Germany, 2001.
25. U. Seiffert and L. Jain, editors. *Self-Organizing Neural Networks: Recent Advances and Applications*, volume 78 of *Studies in Fuzziness and Soft Computing*. Springer-Verlag, Heidelberg, 2001.
26. J. M. Shapiro. Embedded image coding using zerotrees of wavelet coefficients. *IEEE Transactions on Signal Processing*, 41(12):3445–3462, 1993.
27. D. S. Taubman and M. W. Marcellin. *JPEG 2000: Image Compression Fundamentals, Standards and Practice*. Kluwer International, Dordrecht, 2000.
28. V. N. Vapnik. *The Nature of Statistical Learning Theory*. Statistics for Engineering and Information Science. Springer Verlag, Heidelberg, 2. edition, 1999.
29. L. Wang and E. Oja. Image compression by MLP and PCA neural networks. In *Eighth Scandinavian Conference on Image Analysis*, pages 1317–1324, 1993.
30. P. Wayner. *Compression Algorithms for Real Programmers*. Morgan Kaufman, San Francisco, Ca, 1999.
31. M. J. Weinberger, G. Seroussi, and G. Sapiro. LOCO-I: A low complexity, context-based, lossless image compression algorithm. In J. A. Storer and M. Cohn, editors, *Proceedings of the IEEE Data Compression Conference*, pages 140–149, Piscataway, NJ, 1996. IEEE Computer Society Press.
32. X. Wu and N. Menon. Content-based, adaptive, lossless image coding. *IEEE Transactions on Communications*, 45(4):437–444, 1997.

Discriminative Clustering of Yeast Stress Response

Samuel Kaski[1,2], Janne Nikkilä[1], Eerika Savia[1] and Christophe Roos[3]

[1] Neural Networks Research Centre, Helsinki University of Technology
P.O. Box 5400, FIN-02015 HUT, Finland
{samuel.kaski, janne.nikkila, eerika.savia}@hut.fi
[2] Department of Computer Science, University of Helsinki, Finland
[3] Medicel Ltd., Helsinki, Finland
christophe.roos@helsinki.fi

Summary. When a yeast cell is challenged by a rapid change in the conditions, be it temperature, osmolarity, pH, nutrient or other, it starts a genome stress response program. Survival of especially single-cell organisms depends on their ability to adapt to the environmental changes and therefore stress response has received much attention. In the budding yeast *Saccharomyces cerevisiae* several hundred genes out of about 6500 present in the genome have previously been found involved in a stereotyped stress response pattern. Hierarchical clustering techniques applied to gene expression measurements have also previously identified a subset of genes termed common environmental stress response (CESR) or common environmental response (CER) genes, that respond in the same way in a variety of environmental conditions. There is evidence from two different sets of experiments that many of these genes are regulated by the same Msn2p and Msn4p transcription factor pair. We have extended the study by in silico data mining using a new supervised discriminative clustering (DC) technique, which directly searches for responses potentially regulated by the Msn2/4p factors. We observed a cluster of CESR/CER genes, comparable to those previously found and potentially regulated by Msn2/4p. The results of discriminative clustering both support the viability of the technique in supervised gene expression clustering and yield new insights into genomic stress response.

1 Introduction

The ability of the yeast *Saccharomyces cerevisiae* genome to respond to environmental changes is vital, since no cellular condition of gene activity is universally optimal. The response of yeast cells to stress induced by drastic changes in the environment has been used as a paradigm to study gene regulation networks. It is also important to understand the cell response to stress to its own value, since virtually any treatment introduces some kind of stress situation for the yeast cells, and is thus present in any gene activity measurement. Moreover, understanding yeast gene regulation will help as a model for studies on higher organisms. While it is clear that understanding gene regulation requires data on chromosome structure, gene activity (transcriptome), protein pool to mention only the major concepts, the transcriptome has received the most attention due to the high throughput measurement technologies available (gene chip/microarray).

The gene expression of the yeast under stress has been studied extensively [1, 3, 9, 13], and it has become evident that a certain group of the yeast genes is always activated during various stress treatments. The genes in this set are often called *common environmental response (CER)* genes [1], or *environmental stress response, ESR* genes [3]. In this paper we adopt the term from [1], and call them CER genes.

Due to differences in the documentation of the experiments in [3] and [1], it still is somewhat uncertain whether the group of CER genes found in one experimental setting is the same as the set of genes found in the other experiments. Even more unclear is the understanding of the regulatory system of the yeast stress response. There seem to be at least a few general "stress regulators" like Yap1p, Msn2p, and Msn4p, that are shown to be required for a large set of CER genes to be induced [1, 3]. In addition, the existence of condition specific regulators, like Hsf1p for heat shock, has been noted [9].

We carry out a meta-analysis to study the concurrence of the two different CER gene definitions and the two independent sets of measurements. Additionally, we refine *in silico* the earlier analyses of the role of the Msn2p and Msn4p transcription factors in regulating the CER genes. We use a new statistical data mining tool called *discriminative clustering (DC)* [7, 15, 16] which differs from standard clustering by being supervised by class labels of the data.

The clusters in DC partition the data into mutually similar sets, in the same way as in the standard K-means clustering. The difference is that DC maximizes the dependence of the clusters and the classes. An intuitive description of what DC does is that it uses the classes as hints on which samples should be considered similar. Samples should be more similar if they belong to the same class; more precisely, distances in directions where the class distribution changes more should be larger.

In this study the classes are chosen according to the response of the strain lacking Msn2/4p to a stress treatment. Then DC will consider genes more similar if their response is the same even after the potential stress regulator Msn2/4p is removed. The cluster analysis becomes more focused on regulation by Msn2/4p, instead of taking all differences in gene activation into account.

2 Discriminative Clustering

Consider a set of paired data $(\mathbf{x}, c)$, where $\mathbf{x} \in \mathbb{R}^n$ are continuous-valued multivariate observations of primary data and c are discrete classes. In this work each $\mathbf{x}$ is a profile of expression of a yeast gene in various stress treatments. In a nutshell, we wish to find clusters of $\mathbf{x}$ that are maximally dependent on c. This task has two parts. (i) In order to call the data groups *clusters*, they need to be local in the primary data space, that is, contain similar expression profiles. The second part is that (ii) the clustering should capture the dependency between the primary data and the classes.

The motivation for (i) is that even though the clusters are supervised, they can still be interpreted in the same way as "normal clusters" in unsupervised clustering, as sets of similar data. The motivation for (ii) is that choosing the classes properly allows us to focus the analysis to the variation relevant to the classes. In this work we want to find evidence for regulation by Msn2/4p, and we choose the classes to show how the genes react to stress treatments after the Msn2/4p has been removed. Maximization of dependency with the classes then forces the clustering to focus on similarities in the expression profiles that are relevant to regulation by Msn2/4p. Genes regulated in the same way will become more similar.

2.1 Definition of Clusters

Each cluster j is defined by a prototype $\mathbf{m}_j \in \mathbb{R}^n$. Samples $\mathbf{x}$ are assigned to the clusters that have the closest prototype: $\mathbf{x}$ belongs to cluster j if $\|\mathbf{x} - \mathbf{m}_j\| \leq \|\mathbf{x} - \mathbf{m}_k\|$ for all k. Here the distance is the standard Euclidean distance. This definition is the same as in the standard K-means clustering method, for instance.

2.2 Measuring Dependency

The clusters and the classes form a contingency table, a cross tabulation of the two categorizations of the same data. The count of data n_{ji} within cell (j, i) tells how many samples of class i occur in the cluster j. The margin $n_{j\cdot} = \sum_i n_{ji}$ gives the number of samples within cluster j, and the fixed margin $n_{\cdot i} = \sum_j n_{ji}$ gives the total number of samples in class i.

The dependency between the clusters and the classes can be measured based on the contingency table. If the true proportion of data occurring within each cell, i.e. the joint distribution p_{ji}, was known, the dependency could be measured by mutual information. However, since only a finite sample is available, the mutual information computed from the empirical distribution would be a biased estimate. A Bayesian finite-data alternative is the *Bayes factor* between models that assume dependent and independent margins. Bayes factors have classically been used as dependency measures for contingency tables (see, e.g., [4]). We have used the classical results as building blocks to derive the Bayes factor to be optimized; the novelty in DC is that we suggest maximizing the Bayes factor instead of only measuring dependency of fixed tables with it.

2.3 The Cost Function

In general, frequencies over the cells of a contingency table, as well as over the margins, are multinomially distributed. The model M_i of *independent margins* assumes that the multinomial parameters of the contingency table cells are determined by the posterior parameters at the margins. In the alternative model M_d of *dependent margins*, the cell-wise frequencies are assumed to have been sampled directly from a multinomial distribution over the whole contingency table, which indirectly determines the margins. Dirichlet priors are assumed for both the margin and the table-wide multinomials.

Maximization of the Bayes factor

$$\mathrm{BF} = \frac{p(\{n_{ji}\}|M_d)}{p(\{n_{ji}\}|M_i)} \tag{1}$$

with respect to the clusters then gives a contingency table where the margins are maximally dependent, that is, which cannot be explained as a product of independent margins. The cluster margin is determined by the distribution of the learning data set into the clusters, and the clusters in turn are defined by their parameters (the cluster prototypes $\mathbf{m}_j$). The BF is maximized with respect to the parameters.

After marginalization over the multinomial parameters, the Bayes factor, assuming a fixed class margin, takes the form [16]

$$\mathrm{BF} = \frac{\prod_{ji} \Gamma(n_{ji} + n^0)}{\prod_j \Gamma(n_{j\cdot} + N^0)} . \tag{2}$$

Here $n_{j\cdot} = \sum_i n_{ji}$ is the cluster margin, that is, number of data samples in the clusters, and the parameters n^0 and N^0 come from the Dirichlet priors. We have set $n^0 = N^0 = 1$.

For large data sets compared to the number of clusters, (2) is approximated by mutual information of the margins. Another interesting connection, shown in [16], is that the Bayes factor equals the posterior density $p(\{\mathbf{m}\}|D)$ of the set of the cluster parameters $\{\mathbf{m}_j\}$ of a certain predictive model. The model predicts the class distribution within each cluster with a multinomial distribution.

2.4 Optimization

The difficulty in optimizing (2) is that the data counts n_{ji} within the clusters are discontinuous functions of the values of the cluster parameters $\mathbf{m}_j$. The counts change only when a data point changes from one cluster to another. Hence, the derivatives of the cost function are always either zero or undefined.

We have used a heuristic smoothing technique to make gradient-based optimization possible. It has worked about as well as the theoretically better justified simulated annealing that is much heavier computationally [7]. The "number" of samples is smoothed by $n_{ji} = \sum_{c(\mathbf{x})=i} y_j(\mathbf{x})$, where $c(\mathbf{x})$ is the class of $\mathbf{x}$ and $y_j(\mathbf{x})$ is a smoothed cluster "membership function", defined by $y_j(\mathbf{x}) = Z(\mathbf{x})^{-1} \exp(-\|\mathbf{x} - \mathbf{m}_j\|^2/\sigma^2)$ with Z such that $\sum_j y_j(\mathbf{x}) = 1$, and σ governing the degree of smoothing. The standard conjugate gradient algorithm was used for the optimization. The smoothing is used only during optimization; afterwards the clusters partition the data space.

2.5 Related Methods

Discriminative clustering is closely related to the Information Bottleneck principle [2, 17] and distributional clustering [12]. The main difference is that in DC the primary data $\mathbf{x}$ is continuous-valued whereas in distributional clustering it has always been categorical. For continuous-valued data the clusters need to be defined and parameterized as partitions of the data space, which makes the algorithms and solutions very different. Although no algorithm has been developed it could in principle be possible to use the Information Bottleneck definition for continuous data as well. Then the clusters would not be local, however, and hence not as easily interpretable as "normal clusters".

Another line of related work is model-based clustering of the joint distribution of the data [5, 11]. The difference is that DC as such does not model the margin $p(\mathbf{x})$ at all; it is a predictive model of the conditional density $p(c|\mathbf{x})$. The motivation for this choice comes from the learning metrics principle [6, 8] which uses the classes c to derive a Riemannian distance measure to the primary data space. In the new metric the class distribution changes homogeneously, which stretches the directions where the class distribution changes rapidly and contracts the directions where it does not. This is the desired result if changes in the class distribution are the interesting thing in the data. The metric can and has been used to supervise a variety of standard data analysis methods. The connection to DC is that it can be shown [6] that under restrictive assumptions DC is asymptotically equivalent to standard K-means in such a metric.

A direct connection between modeling of joint density and DC is that by including a model for $p(\mathbf{x})$, DC can be regularized to a model of joint density, $p(\mathbf{x}, c) = p(\mathbf{x})p(c|\mathbf{x})$. If $p(c|\mathbf{x})$ comes from standard DC and $p(\mathbf{x})$ from a K-means type model, then the cost function of DC becomes a tunable compromise between K-means and DC [7]. This compromise can be interpreted as regularization of DC towards K-means, which is useful for small data sets, assuming the density structure in $p(\mathbf{x})$ contains useful hints for the prediction task.

3 Data

Both Causton and colleagues [1] and Gasch and colleagues [3] have used DNA microrarrays to analyze changes in the transcriptome (pool of all gene transcripts from a cell or cell population) in yeast cells responding to a panel of diverse environmental stresses. The conditions include treatment with heat, changes in pH, in salt concentration, in osmolarity, in reactive oxygen or nutrient concentrations. In each condition, first a reference time-point is measured and then transcriptome data from a set of consecutive time points following the environmental challenge are gathered. Altogether, each of the two research groups has gathered about 150 microrarray measurements covering the full yeast genome of about 6200 known or predicted genes. Both groups attempt to define "genomic expression programs", in other words groups of genes that are commonly involved in handling most or all stress challenges.

In [1] a set of "Common Environmental Response" (CER) genes is defined as follows: First genes, whose expression was found to be induced or repressed in all conditions, were identified by visual inspection from a hierarchically organized tree. Then the genes that changed at least twofold (up or down) in five or six time courses were selected as CER genes. The authors collected 499 genes with a common response to most of the environmental changes examined. Of the 499 genes, 216 were found up-regulated (activated) while 283 were down-regulated (repressed). In [3] the environmental stress response (ESR) genes are more strictly defined: two hierarchical clusters of genes, one with ca 300 activated, the other with ca 600 repressed genes were identified as having a stereotyped response to each of the stress conditions. In all the time series the data was divided by the value of the respective time point zero.

It is known from previous studies that many stress response genes are under the regulation of the Msn2p and Msn4p transcription factors [10] and therefore both groups also make attempts to measure stress response in yeast strains mutant for these transcription factors. In [1] a CER subset (not documented) of 136 genes is identified based on their opposite behavior in the mutant lacking Msn2/4p as compared to the wild type control in the acid challenge experiment. Since we were not able to obtain this list of 136 genes, we imitated the original preprocessing of the data according to the documentation, resulting in a set of 4146 genes, which we analyze further with DC.

In [3] an ESR subset of 180 genes depending on Msn2/4p or the Yap1p transcription factors is documented. In order to compare these findings with ours we had to find matching genes in the data set of [1]. A match was found for a subset of 143. We will refer to this common set of genes by "dependent CER genes from [3]".

4 Results

4.1 Msn2/4p Regulated CER Genes by Discriminative Clustering

In [1], the CER genes were identified and analyzed in two stages. First, it was assumed that the CER genes react in the same way in all of the stressful environmental conditions. The expression profiles of all genes were clustered, and sets of the most up- and down-regulated genes were identified by visual inspection from the hierarchical clustering tree. Second, the response of the genes to a mutation in the putative regulators, Msn2/4p, was studied. A set of genes up-regulated in the wild type but down-regulated in the mutant strains was identified as CER genes potentially regulated by Msn2/4p.

This setting is perfectly suited for discriminative clustering. The goal is to cluster the expression profiles to discover similarly behaving genes. Yet, pure unsupervised clustering is not enough; it is particularly interesting to search for those similarities in expression that are regulated by Msn2/4p.

In the discriminative clustering setup, the expression profile of a gene is $\mathbf{x}$, and the supervising class label c comes from the response of the gene to the mutation. We quantized the response to acid treatment after mutation (vs. time-point zero) to three classes: **down**: strongly down-regulated after mutation (one quarter of genes); **up**: strongly up-regulated (one quarter of genes); and **no change**: the rest. DC then finds clusters of genes that (1) behave similarly in the set of stress treatments and (2) respond similarly to the mutation.

We start by verifying the technical findings quantitatively, by checking that the dependencies the supervised clustering finds are replicable. Then we interpret the results and compare the findings qualitatively with those in [1]. Due to differences in reporting of the results in the papers, quantitative comparison is possible only with [3]; it will be carried out in Sect. 4.2.

DC Results Are Replicable

In order to verify that the results of supervising the clustering are real and not merely results of overfitting the clusters to noise in the data, we compared them to standard unsupervised K-means clusters with cross-validation.

The smoothing parameter $\sigma = 0.9$ used in the optimization was chosen with a validation set in preliminary experiments, and the number of clusters was set (heuristically) to 12. DC was initialized by K-means.

In the cross-validation study the data was randomly divided into $N = 20$ sets. Clusters were computed with $N - 1$ of the sets, and the results evaluated with the remaining test set. T-test over the $N = 20$ replications showed that the DC consistently ($P < 0.001$) found dependencies between the classes and the expression profiles. The performance measure was (2).

DC Finds a CER Cluster Downregulated in Mutant Lacking Msn2/4p

The expression profiles of the yeast genes in the 12 clusters are shown in **Figs. 1.–3.** DC was computed of the whole set of 4146 genes used in [1]. Each expression profile is a set of time series under different stress treatments. The time labels are shown in **Fig. 4.** and a detailed description of the time series can be found in [1].

The most striking finding is the cluster number 5 (enlarged in **Fig. 4.**), containing 103 genes that have highly upregulated expression in all the treatments for the wild strain and an exceptionally large proportion of the genes are downregulated in the mutant strain lacking Msn2/4p (**Fig. 5.**). This cluster is the most likely candidate for CER genes that are regulated by Msn2/4p.

The possibility that the class distribution within cluster number 5 could have arisen by chance was evaluated by random sampling. If there is no interaction between the classes and the clusters, the distribution of data in the contingency table is determined completely by the distribution of data in its margins, that is, the classes and clusters. We sampled a large set (10,000) of contingency tables under the hypothesis that the margins are independent, and estimated for each contingency table cell how unexpected the observed value is. The P-value for obtaining a more extreme value than the observed number of samples was computed as a percentage within the sampled set.

The resulting P-values for each contingency table cell are shown in **Tables 1.** and **2.**. For cluster number 5 the number of downregulated genes is much larger than expected ($P < 0.001$) and the number of non-affected genes is much lower than expected ($P < 0.001$). Hence, it is very unlikely that the observed interaction of the effect of *msn2/4* mutation and the very active CER-type response profile of the genes would have arisen by chance.

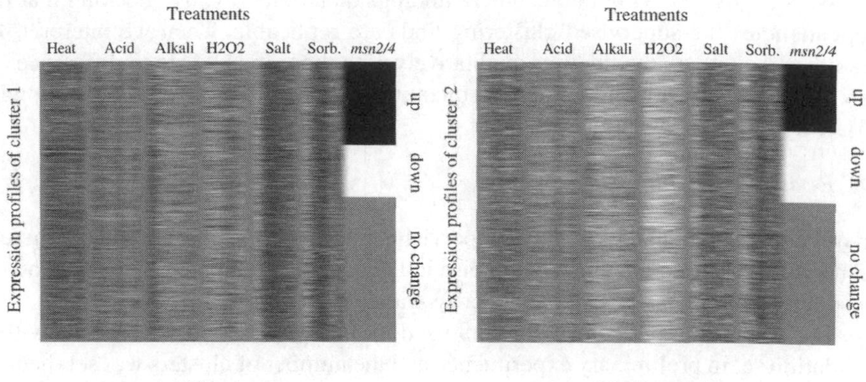

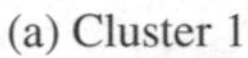
(a) Cluster 1

(b) Cluster 2

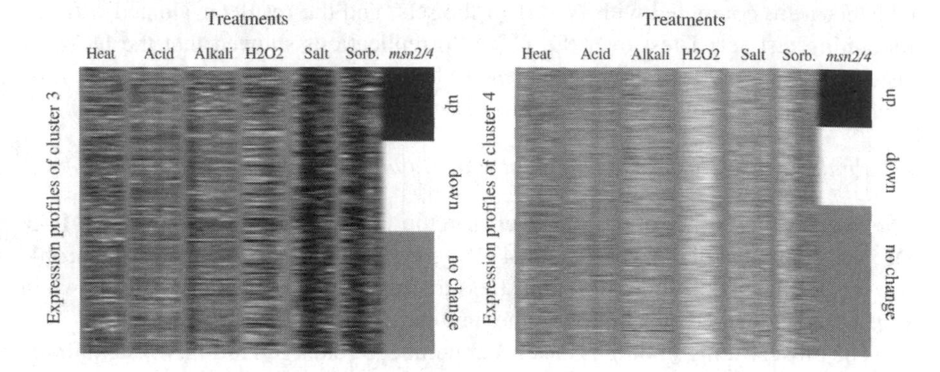

(c) Cluster 3

(d) Cluster 4

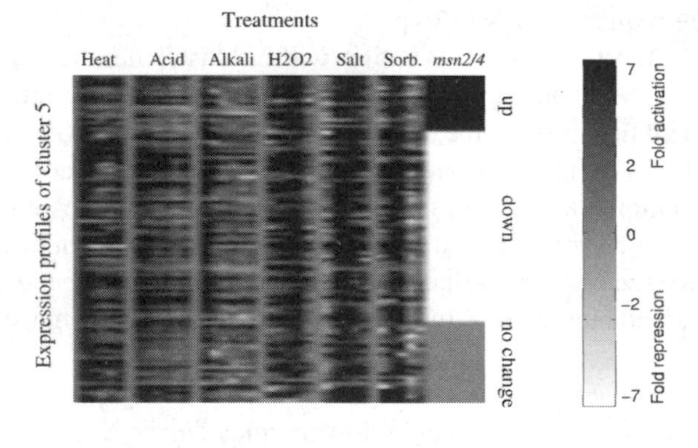

(e) Cluster 5

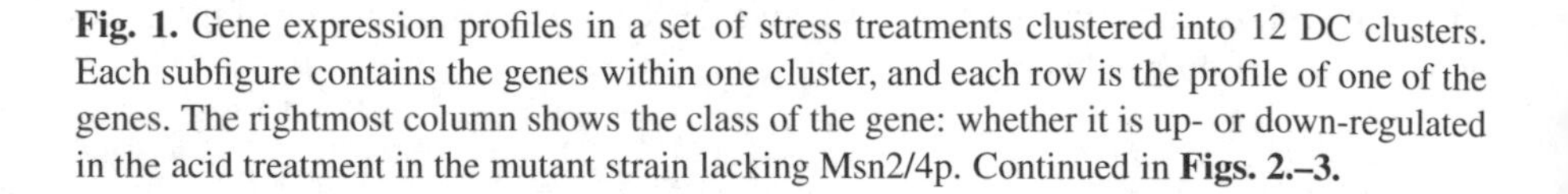
Fig. 1. Gene expression profiles in a set of stress treatments clustered into 12 DC clusters. Each subfigure contains the genes within one cluster, and each row is the profile of one of the genes. The rightmost column shows the class of the gene: whether it is up- or down-regulated in the acid treatment in the mutant strain lacking Msn2/4p. Continued in **Figs. 2.–3.**

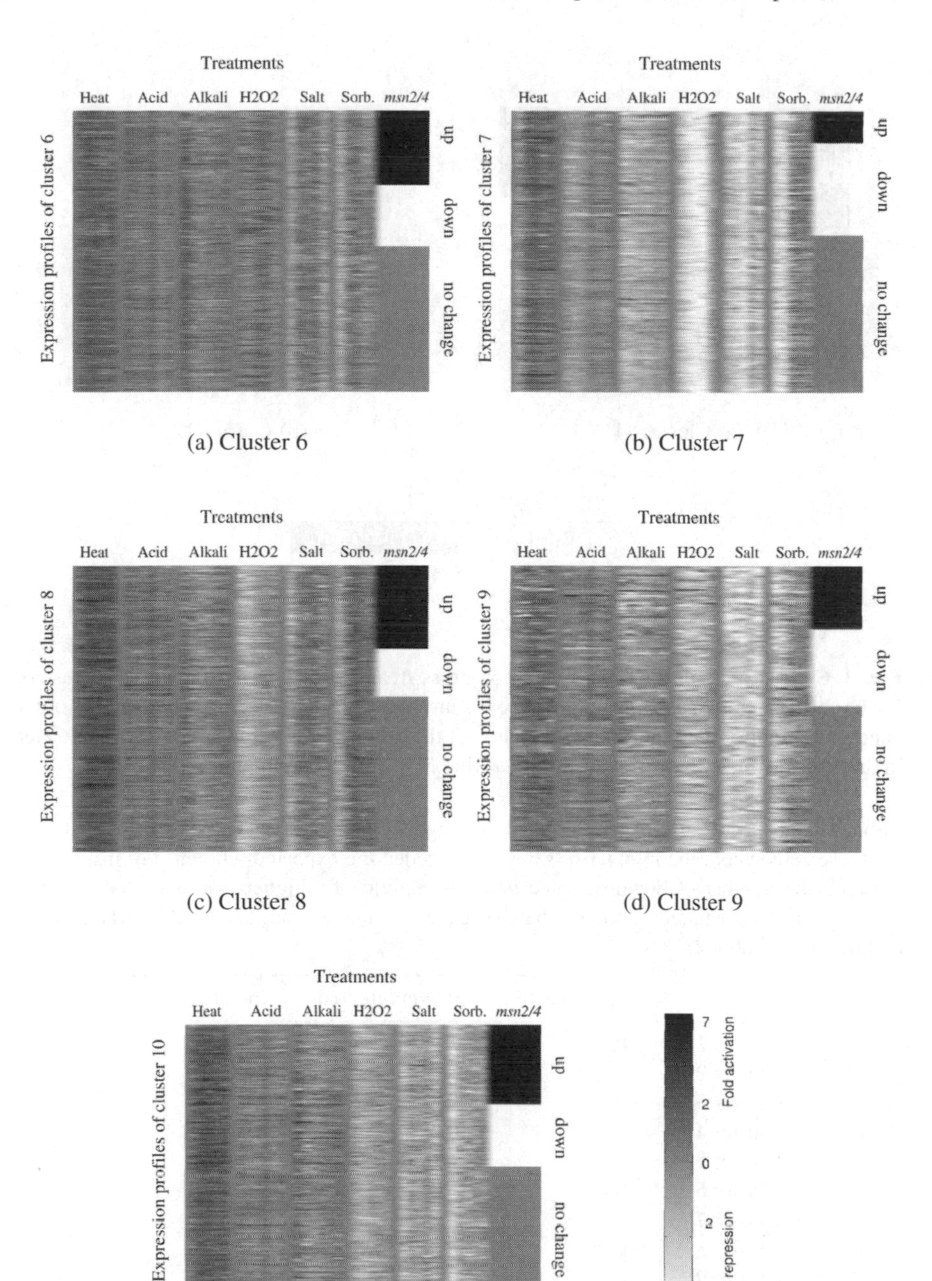

(a) Cluster 6

(b) Cluster 7

(c) Cluster 8

(d) Cluster 9

(e) Cluster 10

Fig. 2. Gene expression profiles in a set of stress treatments clustered into 12 DC clusters. Each subfigure contains the genes within one cluster, and each row is the profile of one of the genes. The rightmost column shows the class of the gene: whether it is up- or down-regulated in the acid treatment in the mutant strain lacking Msn2/4p. Continued in **Figs. 1.** and **3.**

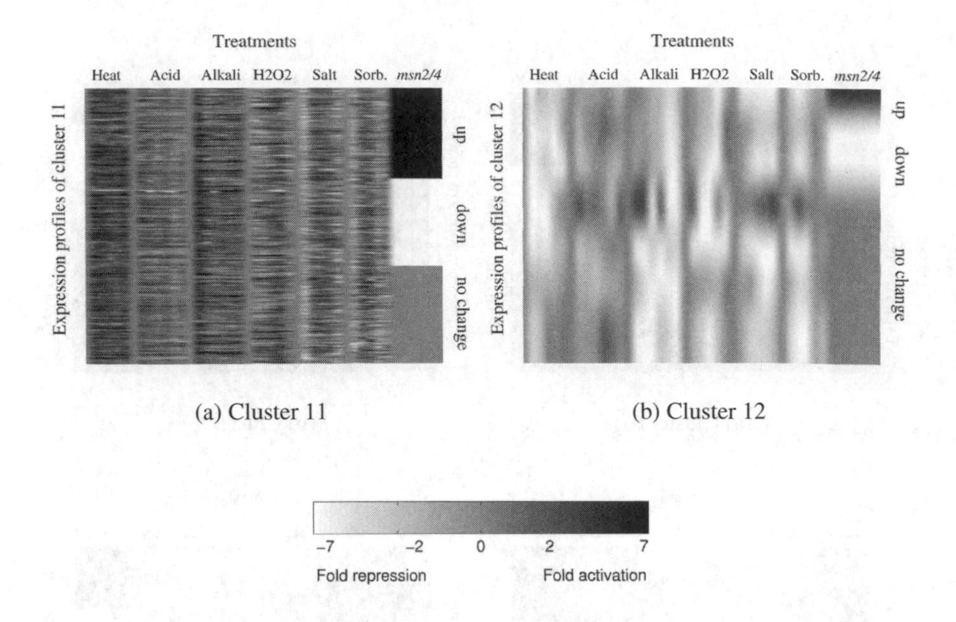

(a) Cluster 11

(b) Cluster 12

Fig. 3. Gene expression profiles in a set of stress treatments clustered into 12 DC clusters. Each subfigure contains the genes within one cluster, and each row is the profile of one of the genes. The rightmost column shows the class of the gene: whether it is up- or down-regulated in the acid treatment in the mutant strain lacking Msn2/4p. Continued from **Figs. 1.–2.**

Table 1. Unexpectedness of the enriched contingency table cells. The table shows P-values for those cells where the number of samples exceeded the expected amount. For instance, in cluster 5 the number of downregulated genes is significantly higher than expected, whereas the number of upregulated and not changed genes is smaller than expected (marked by "–" and treated in **Table 2.**)

	upregulated	downregulated	no change
Cluster 1	0.02	–	0.31
Cluster 2	–	0.41	0.46
Cluster 3	0.40	0.07	–
Cluster 4	–	0.17	0.23
Cluster 5	–	**< 0.01**	–
Cluster 6	0.22	–	0.31
Cluster 7	–	**< 0.01**	0.11
Cluster 8	0.06	–	0.13
Cluster 9	–	0.30	0.43
Cluster 10	0.15	–	0.47
Cluster 11	0.01	0.03	–
Cluster 12	–	0.44	0.45

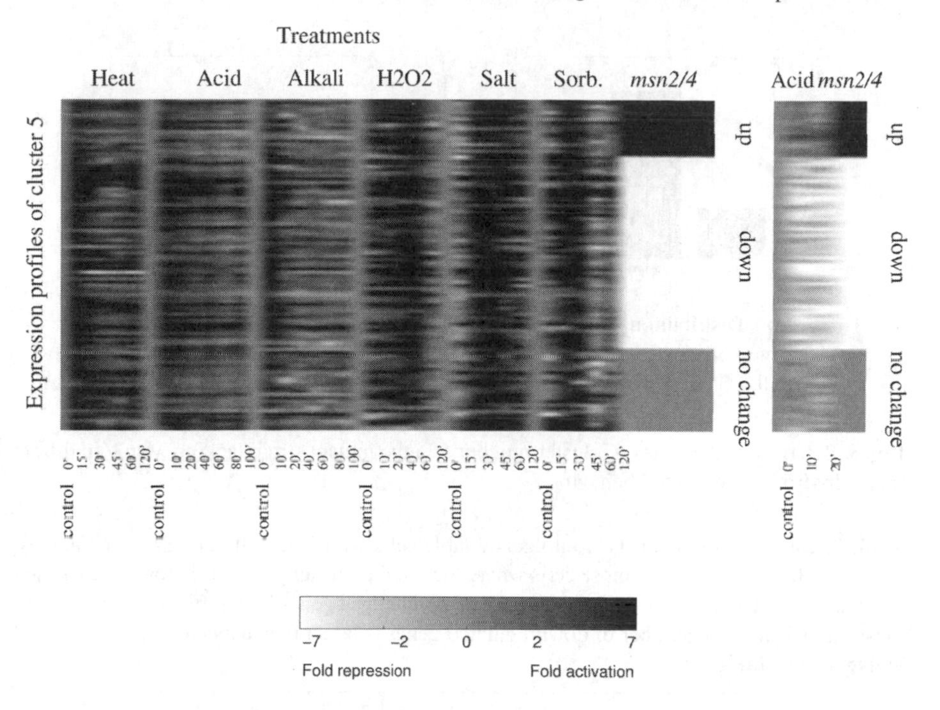

Fig. 4. Left: Enlarged gene expression profiles in a set of stress treatments for genes in cluster 5. Each row represents the expression profile and the class (rightmost column) of one of the genes. The genes have been ordered according to their classification. Right: Expression profiles of the same genes in the mutant strain lacking Msn2/4p. These profiles have been used for defining the classes of the genes (shown in the rightmost columns). The classes tell whether the genes are up- or down-regulated in the acid treatment in the mutant. The order of the genes is the same as in left picture

We cannot verify quantitatively how closely our findings match those of Causton *et al.* [1] since they do not report the full list of gene names. We will, however, compare our list with the list of another study [3] in Sect. 4.2.

Other Findings

We tried to see if the clustering would find a group of stress response genes, the expression of which is independent of the Msn2/4p regulation. Cluster 7 (**Fig. 1 (b)**) contains many down-regulated genes, especially in the peroxide and osmotic shock experiments. Since Msn2/4p are primarily transcription activator factors (and not repressors) [10] these down-regulated genes are probably not under the direct control of Msn2/4p. However, these genes could be under an indirect control of Msn2/4p if one considers that Msn2/4p could activate some secondary repressing regulators. Therefore, the relatively high abundance of down-regulated genes in cluster 7 is not

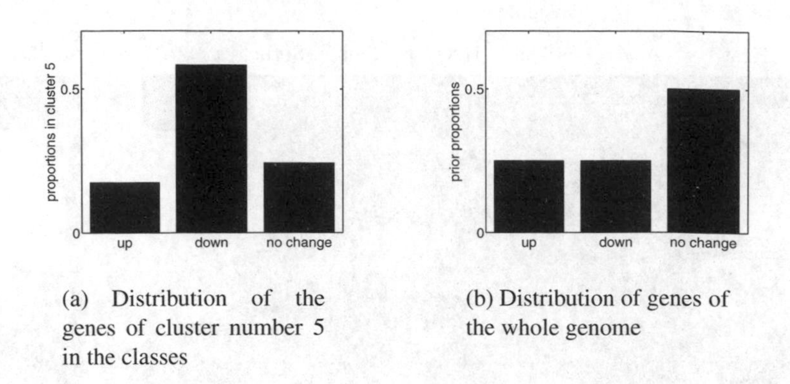

(a) Distribution of the genes of cluster number 5 in the classes

(b) Distribution of genes of the whole genome

Fig. 5. Behavior of the genes of cluster number 5 in the mutant strain lacking Msn2/4p differs markedly from the expected behavior

Table 2. Unexpectedness of the contingency table cells with diminished number of samples. The table shows P-values for those cells where the number of samples falls below the expected amount. For instance, in cluster 5 the number of not changed genes is significantly smaller than expected, whereas the number of downregulated genes is larger than expected (marked by "–" and treated in **Table 1.**)

	upregulated	downregulated	no change
Cluster 1	–	**< 0.01**	–
Cluster 2	0.38	–	–
Cluster 3	–	–	0.11
Cluster 4	0.02	–	–
Cluster 5	0.08	–	**< 0.01**
Cluster 6	–	0.08	–
Cluster 7	**< 0.01**	–	–
Cluster 8	–	**< 0.01**	–
Cluster 9	0.20	–	–
Cluster 10	–	0.13	–
Cluster 11	–	–	**< 0.01**
Cluster 12	0.31	–	–

in itself a reflection of Msn2/4p-independence. Now, if the down-regulation of these genes would be indirectly repressed by Msn2/4p, the expression should rise in the mutant strain lacking Msn2/4p. Interestingly, this does not seem to be the case for a fairly ($p<0.01$, **Table 1**) large amount of these genes that remain down-regulated also in the mutant. Therefore, we conclude that cluster 7 might contain a significant amount of stress response genes independent of the Msn2/4p-regulation.

4.2 DC Findings Are Consistent with Experiments in a Different Stress Treatment

Two groups [1, 3] have sought for yeast stress-induced genes and their regulation by Msn2/4p. The main difference is that the former studied the response of the *msn2/4* mutant in acid stress and the latter in hydrogen peroxide and heat stress. The independent sets of measurements were made with different measurement techniques (cDNA microarrays vs. Affymetrix chips). If the genes are true CER genes they should of course react generally to any type of stress, and hence be equally detectable in either set of experiments.

So far in this article we have only used the measurements of one of the groups [1]. Now the results of the other group will be used in an independent evaluation to verify our findings. Since replication studies are relatively scarce in large-scale gene expression studies because of the cost of the measurements, it will additionally be interesting to see how consistent the findings from the two data sets are. Our study provides some indirect evidence on this.

The Findings Are Consistent

As a sanity check, we first compared whether the set of CER-type genes found to be down-regulated in the *msn2/4* mutants in the independent study [3] were down-regulated in [1] as well. A matching gene was found for a subset of 143 genes; we will refer to this set as "dependent CER genes from [3]". Within this set, exceptionally many genes are down-regulated in the independent measurements of the acid treatment [1] as well (**Fig. 6.**). The distribution differs significantly (chi-square test, $P < 0.001$) from the expected distribution estimated from the whole data set, which completes our sanity check.

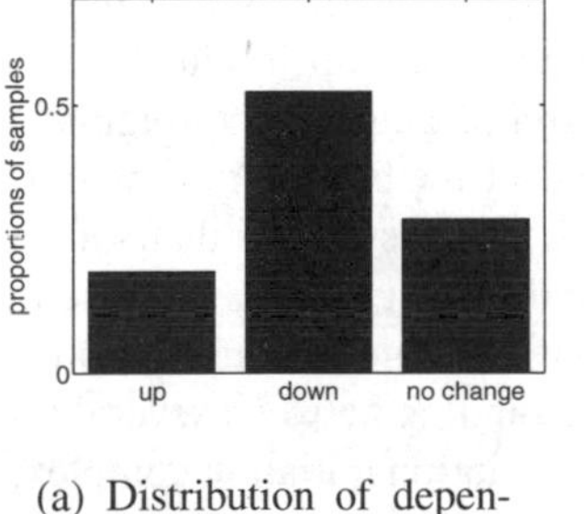

(a) Distribution of dependent CER genes from [3]

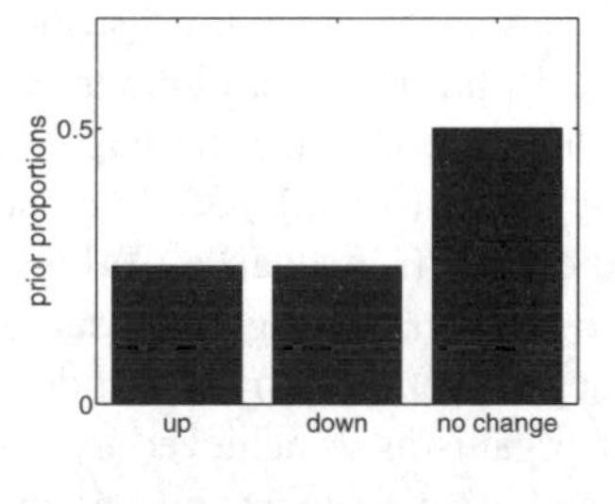

(b) Expected distribution

Fig. 6. Behavior of dependent CER genes from [3] in the acid treatment of [1] (**a**) differs strongly from the overall expected behavior computed from all the genes in [1] (**b**)

Next, we used the results of the independent study to verify the DC results. Based on **Fig. 5**, the cluster number 5 should contain a large proportion of CER genes regulated by Msn2/4p. This finding is based on analyzing one of the data sets with DC, and now the result is compared with a non-DC analysis of the other independent set. If the result is favourable, it will support the viability of DC.

Fig. 7. shows that the proportion of the independently found CER genes in cluster number 5 is exceptionally high; the number differs significantly from chance (chi-square test, $P < 0.001$).

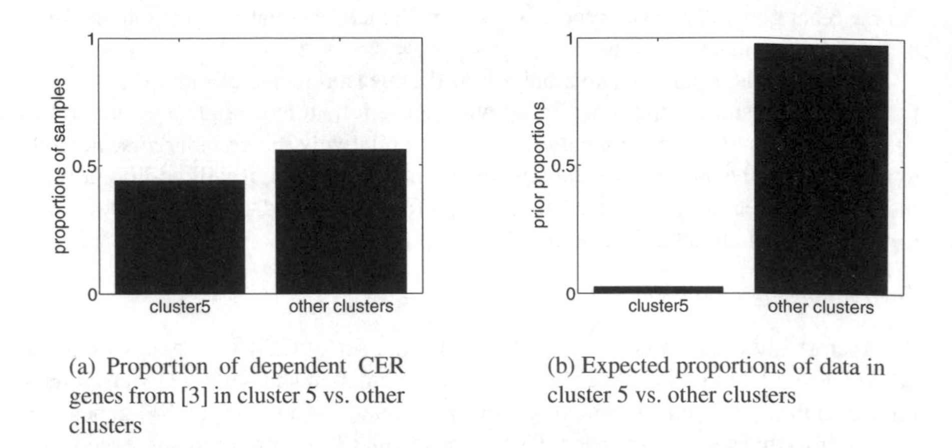

(a) Proportion of dependent CER genes from [3] in cluster 5 vs. other clusters

(b) Expected proportions of data in cluster 5 vs. other clusters

Fig. 7. The dependent CER genes found in an independent study [3] are considerably enriched in cluster 5 (**a**), compared to the expected number of genes calculated from all the genes in [1] (**b**)

Not all of the dependent CER genes found in [3] belong to cluster 5, however. Nevertheless, they are distributed very inhomogeneously in the DC clusters (**Fig. 8.**). In particular, a number of them have ended up in clusters 1, 3, and 11. These clusters contained the largest proportion of generally up-regulated genes in the DC-clusters (**Figs. 1.–3.**). In clusters 1 and 11 the behavior of the mutant strains differed clearly from chance (**Tables 1.** and **2.**). This suggests that some genes predicted to be Msn2/4p-regulated end up in the different clusters because they are not solely dependent on Msn2/4p. Indeed, regulation of gene transcription in yeast, as in other organisms, is achieved by synergistic binding between several transcription factors and other proteins building up the transcription initiation complex.

5 Discussion

In summary, we have applied a new supervised clustering method, discriminative clustering (DC), to mine gene expression profiles for common environmental response (CER) genes and their regulatory mechanisms.

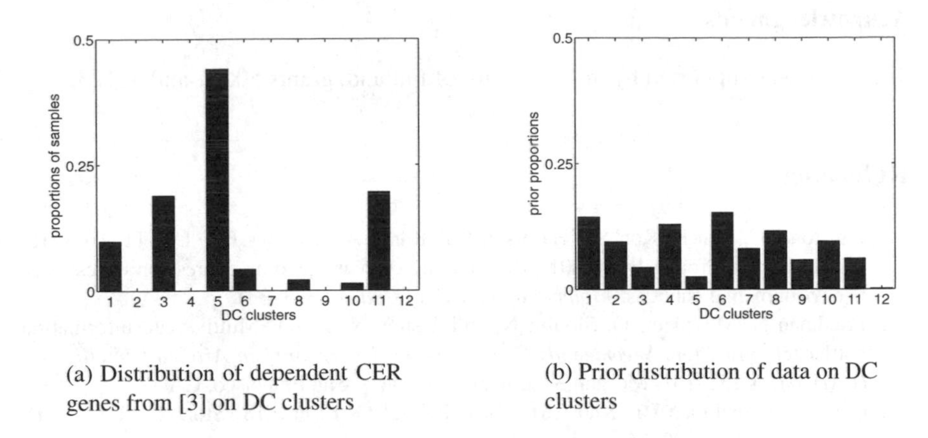

(a) Distribution of dependent CER genes from [3] on DC clusters

(b) Prior distribution of data on DC clusters

Fig. 8. The dependent CER genes found in an independent study [3] are concentrated on only a few DC clusters (**a**), compared to the distribution expected based on the whole data from [1] (**b**)

The clustering was supervised to focus on gene expression relevant to regulation by certain transcription factors, Msn2/4p. The findings are consistent with both of the two earlier studies on the same problem [1, 3]. DC has been originally developed for supervised mining of large data sets, and the results support its usefulness in genome-wide mining of expression data.

Additionally, the DC clustering suggested possible subclasses within the set of CER genes.

The DC complements standard unsupervised clustering by making it possible to supervise the exploration of data. Ultimately, when the resulting hypotheses mature, they need to be tested with even more focused methods and models. The current findings suggest a follow-up study where the stress-induced genes that are down-regulated in mutants lacking *Msn2/4p* mutants would be sought directly by searching for genes with high activity (up-regulation) in the wild type and low activity (down-regulation) in the mutants.

As a side study, we compared indirectly the results of two research groups, working with different methods and published in different papers [1, 3]. To the extent the documentation allows, the results seemed compatible. It would be interesting to continue the present DC study by generalizing from one supervisory signal, class labels, to multiple classifications derived from the response of the mutant strains to different stress treatments. Data is already available by [3]. The results should reveal more about the compatibility of the different data sets and should yield more accurate hypotheses about which genes are true CER-genes and respond similarly to all kinds of stress treatments. Moreover, instead of quantizing the responses to three classes they could be considered as multivariate continuous-valued observations. Then the recent generalization of discriminative clustering from categorical supervisory signal to continuous-valued multivariate signal [14] could be the proper data analysis tool.

Acknowledgments

This work was supported by the Academy of Finland, grants 50061 and 52123.

References

1. Causton HC, Ren B, Koh SS, Harbison CT, Kanin A, Jennings EG, Lee TI, True HL, Lander ES, and Young RA (2001) Remodeling of yeast genome expression in response to environmental changes. *Molecular Biology of Cell*, 12:323–337
2. Friedman N, Mosenzon O, Slonim N, and Tishby N (2001) Multivariate information bottleneck. In *Proc. Seventeenth Conference on Uncertainty in Artificial Intelligence (UAI)*, pages 152–161. Morgan Kaufmann Publishers, San Francisco, CA
3. Gasch AP, Spellman PT, Kao CM, Carmel-Harel O, Eisen MB, Storz G, Botstein D, and Brown PO (2000) Genomic expression programs in the response of yeast cells to environmental changes. *Molecular Biology of the Cell*, 11:4241–4257
4. Good IJ (1976) On the application of symmetric Dirichlet distributions and their mixtures to contingency tables. *Annals of Statistics*, 4:1159–1189
5. Hastie T and Tibshirani R (1996) Discriminant analysis by Gaussian mixtures. *Journal of the Royal Statistical Society B*, 58:155–176
6. Kaski S and Sinkkonen J (2002) Principle of learning metrics for data analysis. *The Journal of VLSI Signal Processing-Systems for Signal, Image, and Video Technology, Special Issue on Data Mining and Biomedical Applications of Neural Networks*, accepted for publication.
7. Kaski S, Sinkkonen J, and Klami A (2003) Regularized discriminative clustering. In Molina T, Adali T, Larsen J, Van Hulle M, Douglas S, and Rouat J (eds), *Neural Networks for Signal Processing XIII*, pages 289–298. IEEE, New York, NY
8. Kaski S, Sinkkonen J, and Peltonen J (2001) Bankruptcy analysis with self-organizing maps in learning metrics. *IEEE Transactions on Neural Networks*, 12:936–947
9. Mager WH and De Kruijiff AJ (1995) Stress-induced transcriptional activation. *Microbiological Reviews*, 59:506–531
10. Martinez-Pastor MT, Marchler G, Schuller C, Marchler-Bauer A, Ruis H, and Estruch F (1996) The saccharomyces cerevisiae zinc finger proteins Msn2p and Msn4p are required for transcriptional induction through the stress response element (STRE). *EMBO Journal*, 15:2227–2235
11. Miller DJ and Uyar HS (1997) A mixture of experts classifier with learning based on both labelled and unlabelled data. In Mozer M, Jordan M, and Petsche T, (eds), *Advances in Neural Information Processing Systems 9*, pages 571–577. MIT Press, Cambridge, MA
12. Pereira F, Tishby N, and Lee L (1993) Distributional clustering of English words. In *Proceedings of the 30th Annual Meeting of the Association for Computational Linguistics*, pages 183–190. ACL, Columbus, OH
13. Ruis H and Schuller C (1995) Stress signaling in yeast. *Bioessays*, 17:959–965
14. Sinkkonen J, Nikkilä J, Lahti L, and Kaski S (2003) Associative clustering by maximizing a bayes factor. Technical Report A68, Helsinki University of Technology, Laboratory of Computer and Information Science, Espoo, Finland
15. Sinkkonen J and Kaski S (2002) Clustering based on conditional distributions in an auxiliary space. *Neural Computation*, 14:217–239

16. Sinkkonen J, Kaski S, and Nikkilä J (2002) Discriminative clustering: Optimal contingency tables by learning metrics. In Elomaa T, Mannila H, and Toivonen H (eds), *Proceedings of ECML'02, 13th European Conference on Machine Learning*, pages 418–430, Berlin, Springer
17. Tishby N, Pereira FC, and Bialek W (1999) The information bottleneck method. In *37th Annual Allerton Conference on Communication, Control, and Computing*, pages 368–377, Urbana, Illinois

A Dynamic Model of Gene Regulatory Networks Based on Inertia Principle

Florence d'Alché–Buc[1], Pierre-Jean Lahaye[1], Bruno-Edouard Perrin[1,2], Liva Ralaivola[1], Todor Vujasinovic[2], Aurélien Mazurie[2], and Samuele Bottani[2]

[1] Laboratoire d'Informatique de Paris 6, CNRS UMR 7606, 8 rue du capitaine Scott, 75015 Paris, FRANCE florence.dalche@lip6.fr
[2] Laboratoire de Génétique Moléculaire de la Neurotransmission et des Processus Dégénératifs,CNRS UMR 7091, Hôpital La Pitié-Salpêtrière, 75013 Paris, FRANCE

1 Introduction

In molecular biology, functions are produced by a set of macromolecules that interact at different levels. Genes and their products, proteins, participate to regulatory networks that control the response of the cell to external input signals. One of the most important challenge to biologists is undoubtedly to understand the mechanisms that govern this regulation, and to identify among a set of genes which play a regulator role and which are regulated. While the problem used to be approached by a gene to gene approach, this is changed significantly by the development of microarray technology. Expression of thousands of genes of a given organism or a given tissue can now be measured simultaneously on the same chip. This revolution opens a large avenue for research on reconstruction of gene regulatory networks from experimental data.

In this chapter, we claim that both machine learning and modeling of dynamical processes offer a formal and methodological framework to tackle this problem. In our approach, gene regulatory networks are considered as complex, distributed and dynamic systems. The first step consists in the definition of a model of the underlying dynamics. Then, the availability of gene expression kinetics makes it possible to learn parameters of the model. A major advantage of considering the dynamics of the system lies in the fact that the identification step yields both the interaction graph between genes and a simulator for the system. We propose here a linear dynamical model which captures complex behaviours of interactions and a machine learning scheme to identify its parameters. The framework of linear Gaussian state-space models provides a way to take into account noise both in the observations and in the underlying dynamical process of regulation. We then adapt the well known EM algorithm to our specific model.

We illustrate and evaluate the approach on experimental data that concern the SOS DNA repair network of *Escherichia coli*. While this validation has yet to be extended to other kinds of networks, the results highlight some powerful features of our method such as the introduction of inertia and the management of missing values and noise.

The chapter is organized as follows. The problem of reverse engineering of gene regulatory network is described in the second section. Then, in the third section, we introduce an original model based on second order differential equations that handles the presence of inertia. The fourth section is dedicated to learning parameters with a discussion about the merits of two computational intelligence paradigms. Section 5 relates the numerical results we obtained with expression data in *Escherichia coli* within the dynamic bayesian framework. Finally, we present a conclusion and elaborate some future directions in the last section.

2 Elucidating gene regulatory networks from data

Elucidating gene regulatory networks from data can be expressed as a reverse modeling problem. While direct modeling has been developed during the two last decades, especially with the approach of Thomas[23], indirect modeling results from the recent development of high-throughput DNA chips technologies. We would like to extract from observation of gene expressions the structural and functional parameters of an appropriate model. Some approaches consider gene network as static systems and thus, search for the best structure of a network. This network can be deterministic or probabilistic such as the bayesian networks of [6]. We will not consider here static models, focusing on the approaches that allow simulation of the dynamic behavior of the network once it has been modeled.

Among the existing approaches, one major trend consists in starting from a model of the underlying dynamics of gene interactions [2] and then to apply machine learning tools. The most widely used models are additive [5, 14]. Under this term are gathered models which determine the expression of a gene by using a ponderated sum of all expression levels of the others genes. The simpler among these models [5] is purely linear and does not allow to extract non-linear interactions in the network, but can bring to light the most evident relations. A saturation function can optionally be added to avoid divergences and to make the learning easier.[24] described all models of this family [14], [5],[26] in a unified framework of generalized differential equations:

$$\frac{dE_i^t}{dt} = R_i.g(\sum_j w_{ij} E_j(t) + \sum_k v_{ik} z_k(t) + B_i) - \lambda_i E_i(t) \qquad (1)$$

with the following meaning of notations:

- $g(.)$: monotonic regulation-expression function
- $E_i(t)$: expression of gene i at time t
- w_{ij}: strength of regulation of gene j on gene i
- $z_k(t)$: k-th external input on gene i
- v_{ik}: influence of the k-th external input on gene i
- B_i: basal expression of gene i

- λ_i: degradation constant of the i-th gene expression product

These models have been used with different real datasets by [5], [14],[26] and extensively tested on artificial data by [24]. This deterministic framework corresponds to classical artificial recurrent networks.

Conversely, bayesian networks can model the expression of each gene as a conditional probability function of the expressions of the other genes. They are therefore well suited for learning from noisy data. Some well-known algorithms for learning Bayesian networks exist [10], and new algorithms for learning very complex models have recently been proposed [7]. One can find in [15] a theoretical point of view of learning dynamica bayesian networks in the context of gene expression data but results on real experimental data come from [3] that added noise to the additive models presented in 1. [16] have also used dynamic Bayesian networks associated with boolean variables to model regulatory pathways in *E.coli*. An algorithm which identifies interaction networks from dynamic Bayesian networks coupled with a non-parametric regression method has also recently been proposed by [12]. Its main advantage lies with the possibility to reduce noise by using smoothed profiles models of kinetics. However the introduction of prior knowledge in this approach does not seem to be easy.

In the following section, we introduce a new additive model of gene regulatory networks that allow to capture delays and inertia. We will then discuss how to tackle the problem of its identification.

3 Modeling gene interactions using an inertia principle

Let us consider two genes A and B, A regulating positively B. If we observe the evolution of their expression levels,we notice that when gene A begins to be expressed at time t_1, then gene B will be expressed at time $t_1 + \delta t$. On the contrary, if gene A begins to be no more expressed at time t_2, there is a delay before gene B is itself no more expressed as illustrated in the artificial profiles of gene expression in figure1. This delay can also be visualized in figure 2 by suppressing time and representing gene B expression according to gene A. We can see an hysteresis that expresses the underlying inertia of the system.

In order to reflect the presence of inertia, we propose to model the two genes as coupled dampened oscillators. Such an assumption allows to model complex behaviours of gene expression while keeping the differential equations linear. Indeed, only second-order linear differential equations are needed to translate this inertia. If we consider only the regulation of gene A on gene B, we get the following equations:

$$\frac{d^2E_B(t)}{dt^2} + 2\lambda_B\omega_B\frac{dE_B(t)}{dt} + \omega_B^2E_B(t) = w_{BA}E_A(t) \tag{2}$$

Of course, it should be emphasized that the model is able to capture an unoscillating behaviour of gene B if ω_B is sufficiently low and λ_B is high enough. We discuss here of a plausible interpretation of the parameters used in these equations:

Fig. 1. Gene A up-regulates gene B

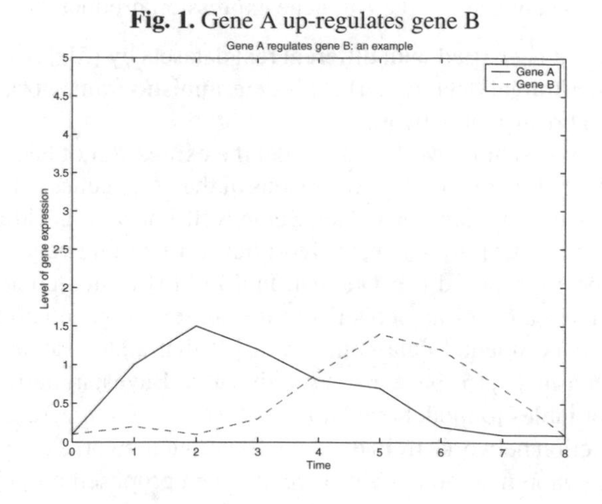

Fig. 2. Hysteresis phenomenon when gene A up-regulates gene B

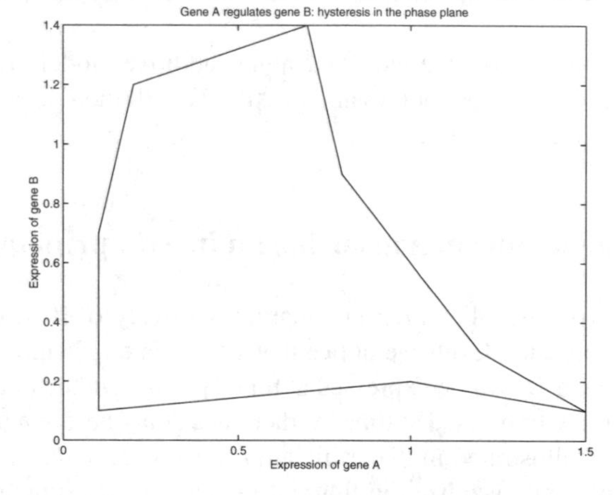

- λ_B is a coefficient without dimension representing the absorption of the gene B. It must be greater than 1 to avoid oscillations. It determines the time necessary to go back to steady state after an excitation, i.e. the time after which no regulated mRNA survives the regulator mRNA. It may be due to the unobserved remanent regulator protein.
- ω_B is the natural frequency of gene B, a positive quantity. It defines a characteristic response to an excitation. It takes into account the time delay necessary for transcription and translation to begin.
- w_{BA} is the coupling strength defining the interaction, positive (excitation) or negative (inhibition), exercised by gene A over gene B. $w_{BA} = 0$ indicates that gene A does not influence gene B. It is important to notice that w_{BA} and w_{AB} in the symmetrical equation have no reason to be equal.

The model can be generalized to a network composed of n genes. $E_i(t), i = 1...n$ is the expression level of gene i at time t. Furthermore, we assume that the model is additive e.g. regulatory genes have a linear cumulative effect on a regulated gene. The system is therefore governed by the following system of n second order differential equations:

$$\frac{d^2E_i(t)}{dt^2} + 2\lambda_i\omega_i\frac{dE_i(t)}{dt} + \omega_i^2 E_i(t) = \sum_j w_{ij}E_j(t), \quad \lambda_i \geq 1. \tag{3}$$

If we introduce $L_i(t)$ as the derivative of expression level for each gene i defined by $L_i(t) = \frac{dE_i(t)}{dt}$, we can now transform the system (3) of n second order differential equations to an equivalent system of $2n$ first order equations:

$$\begin{cases} \frac{dE_i(t)}{dt} = L_i(t) \\ \frac{dL_i(t)}{dt} = -\omega_i^2 E_i(t) - 2\lambda_i\omega_i L_i(t) + \sum_j w_{ij}E_j(t) \end{cases} \tag{4}$$

This system of $2n$ equations can be discretized to be numerically implemented. Assuming that the measurement time unit is lower than the gene evolution characteristic time, we shall consider that continuous derivatives $L_i(t)$ are replaced by: $\frac{\Delta E_i(t)}{\Delta t} = E_i(t+1) - E_i(t)$. We also shall define the vector $\mathbf{x}_t$ as the state vector defined by :

$$\mathbf{x}_t = \left(E_1(t), \cdots, E_n(t), \frac{\Delta E_1(t)}{\Delta t}, \cdots, \frac{\Delta E_n(t)}{\Delta t} \right)' \tag{5}$$

.

Using these notations, the evolution of the network of n genes can therefore be described by the following equation

$$\mathbf{x}_{t+1} = A.\mathbf{x}_t \tag{6}$$

with matrix

$$A = \begin{bmatrix} I & I \\ W - \Omega^2 & I - 2\Omega\Lambda \end{bmatrix} \tag{7}$$

where I is the identity matrix of size $n \times n$, $W = (w_{ij})_{1 \leq i,j \leq n}$, $\Omega =$ diag $(\omega_1, \ldots, \omega_n)$ and $\Lambda =$ diag $(\lambda_1, \ldots, \lambda_n)$ [3].

The linearity of the system provides two advantages:

- stability can easily be analysed [4]
- the identification process will be simpler than for a non linear system.

---

[3] In this chapter diag $(d_1, \ldots, d_n)$ is the diagonal matrix of size $n \times n$ whose diagonal elements are $d_1, \ldots, d_n$ and the others zero; for a square matrix M, diag (M) is the diagonal matrix whose diagonal elements are the diagonal elements of M.

[4] Let us recall that the linear system described by equation 6 is stable if and only if the eigenvalues of the matrix Λ are of magnitude equal or inferior to 1.

4 Learning parameters of the inertial model

4.1 Introduction to machine learning

Machine learning aims at inferring mathematical models from examples in order to solve complex tasks such as pattern recognition and data-mining. In this framework, a mathematical model refers to a probability law, a set of rules or a function chosen in a given family. The learning process is considered as successful if the learned system can deal correctly with new data that come from the same parent probability law. This ability to behave correctly in front of new data is called the *generalization property*. Since 1985, machine learning have developed both theoretical and empirical tools to ensure that learned models generalize well.

In the following we first consider the whole learning process through a general methodology that includes 3 main steps: representation, optimization and validation. At least the two first steps are somehow interdependent and require prior knowledge to be processed. The combined use of data and knowledge is characteristic of artificial intelligence and machine inference. We think that such an approach is especially well-suited to a modelling process and we show it in the context of the reverse modelling problem. Hence, reverse engineering of gene regulatory networks can be formalized as follows:

Let us note $S = \mathbf{x}_1, ..., \mathbf{x}_T$, the observed time-series describing the state of the n considered genes. We assume that H is a given family of functions such that $h_\theta(\mathbf{x}_t)$ can be used to approximate $\mathbf{x}_{t+1}$. θ is the set of parameters of h. We also refer to K as the set of prior available knowledge including the choice of H and optionally some information about the nature of the interactions. A learning algorithm a is a function of data S and knowledge K and it procides an estimation $\hat{\theta}$ of optimal parameter with respect to the maximization of the generalization property.

$$\hat{\theta} = a(S, K) \tag{8}$$

The methodology of machine learning to define the algorithm a is illustrated in the figure 3. The conception of the algorithm a requires first to decide how to represent the data, and which model to implement. This is the *representation step*. We also need to choose according what criterion the parameters θ have to be optimized and how this can be performed: this is the *optimization step*. Lastly, the results have to be evaluated and validated either in an automatic way or with the help of the biologist: this is the *validation step*.

Let us detail each of these issues:

- **the representation issue**
  We must decide how to encode the available data and which class of models we need to fit them. In our problem, the raw data are kinetics of mRNA concentration for each gene. We have to decide if we work directly on these variables with continuous values or if we discretize them. We can also extract some features for instance with signal transformations and use them as new variables. In this case, preprocessing the data makes the role of the model easier.

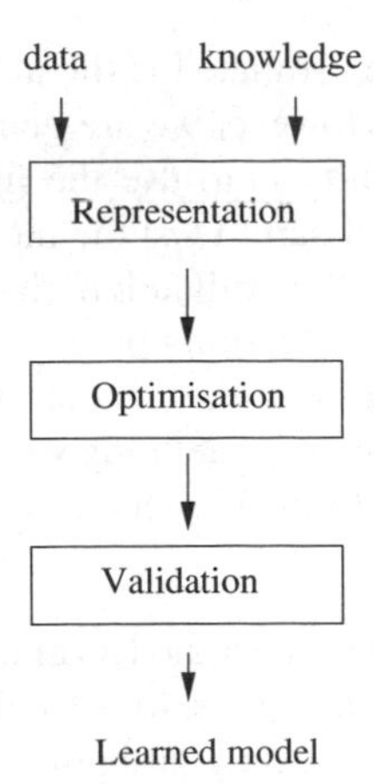

Fig. 3. Machine learning methodology

Choosing a family of models requires to take into account two important properties : one one hand, the class of models has to be sufficiently expressive to represent accurately the system to be modelled and on other hand, each model needs to be flexible in order to get its complexity controlled. For function approximation and time-series modelling, the control of the norm of the parameters allows usually to keep the model simple.

- **the optimization issue**

  Learning has to be set as an optimization problem where objective function is clearly defined. The objective function to be optimized must reflect the goal of the learning process which is to get good generalization properties. Several theories from regularization one [8] to those of large deviations theorems [25] prone to define a cost function that makes the balance between empirical error and sparsity of the model. The model is thus required to fit the available data while keeping simple.

  When the optimization problem is established, an optimization algorithm is developed to determine the parameters of the given model. According to the nature of the problem to be solved, mathematical programming, gradient descent, Expectation-Maximization algorithms or evolutionary algorithms can then be used. It should be emphasized that the relevance of the learning results depends not only on the appropriateness of the cost function but also of the convergence properties of the optimization algorithm. Some algorithms are proven to fall into local minima of the cost function and this has to be taken into account when exploiting the results.
- **the validation issue**

  As learning is performed using a finite sample of data, it must be put into question. Once of criterion is defined to evaluate the quality of the result,robust methods of statistical estimation such as bootstrap or cross-validation can be used. Within the context of time-series analysis and dynamic modelling, the classic

criterion which can be estimated is the ability to make predictions, for instance one of k steps ahead. However we are generally not provided with several time-series and one possibility is to use the first part of the time-series to perform learning and the second part to test the model.
Yet there is another kind of validation: the validation by the domain expert who is here the biologist. The learning process provides an hypothetic model of gene interactions that the biologist must check either by comparing the obtained model with existing true model or by defining some experiments that will prove or refute the hypothetic model. In the numerical results we present, we used the two first kinds of validation.

After this introduction to machine learning principles, we now describe how we have exploited biological knowledge to solve the representation and the optimization questions for the reverse modelling problem.

4.2 Solving the representation issue using the inertial model

In the following, we consider that the data encoding is very simple : we keep kinetics of gene expression with continuous values and do not discretize them in order to avoid quantization noise. The choice of the models family is now reduced to the implementation of the inertial model in a framework that favors learning.

Deterministic implementation as a recurrent artificial neural network

The linear dynamic model that we proposed belongs to the family of linear recurrent artificial neural networks but with a specificity: the equations we have derived describe a distributed system composed of two kind of units. Some units that can be associated with genes which correspond to the E_i variables and some others are latent units which correspond to the L_i variables. Information propagates from t to $t+1$. So the network can be described as recurrent since output at time $t+1$ become inputs to compute the state of the network at time $t+2$ and so on. Within this framework, learning could be performed using backpropagation through time algorithm (BPT) with a possible non linearization of the units to improve the learning process. However we identified one major drawback for this approach: it does not handle noise specifically and moreover, it does not address the missing variables problem which is a frequent case in biological settings.

Probabilistic implementation as a linear Gaussian state-space model

Another the idea is to add noise to the inertial model and consider it as a linear Gaussian state-space model. First we assume that the true process modeled by the equation 6 is hidden and only accessible by the observations that it produces. Second we propose to take into account two sources of noise: the intrinsic noise which is known to be present in genetic expression [13] and measurement noise which is due to the acquisition technologies.

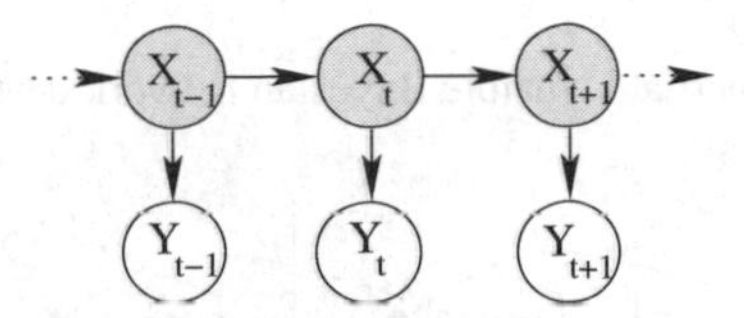

Fig. 4. Linear dynamical system. Dark nodes are hidden. Bright nodes are observed.

In this work, we have made the simplest assumptions about noises and have focused on the gaussian hypothesis, either for intrinsic or measurement noise.

Using the linear dynamical system formalism (cf. Figure 4), these assumptions lead to the following model:

$$\begin{cases} \mathbf{x}_{t+1} = A.\mathbf{x}_t + \mathbf{u} \\ \mathbf{y}_t \quad = C.\mathbf{x}_t + \boldsymbol{\mu}_{obs} + \mathbf{v} \end{cases} \tag{9}$$

$\mathbf{x}_t$ is the hidden state of the gene network at instant t (cf. Equation 5), while $\mathbf{y}_t$ is the observed state of the network, composed of all observations of gene expression levels.

A is the transition matrix of Equation 7 and C the projection matrix $[I \quad \mathbf{0_{n,n}}]$ of size $n \times 2n$, $\mathbf{0_{m,n}}$ being the zero matrix of size $m \times n$, and $\boldsymbol{\mu}_{obs}$ a measurement adjustement vector. Elements $\mathbf{u}$ and $\mathbf{v}$ are independant and identically distributed (i.i.d.) realizations of two Gaussian random variables with zero mean and variances σ_x^2 et σ_{obs}^2. $\mathbf{u}$ and $\mathbf{v}$ express the fact that both biological and measurement phenomena are stochastic. Variables $\mathbf{x}_t$ are usually said to be *hidden* because they only are accessible indirectly through observation of $\mathbf{y}_t$. We make the hypothesis that $\mathbf{x}_1$ follows a Gaussian law of mean $\boldsymbol{\mu}_i$ and variance σ_i^2.

The proposed model (9) can be also be described explicitely using conditional probabilities and can be seen as a *dynamic Bayesian network* (DBN) with hidden nodes (cf. Fig.5)

$$\begin{cases} p(\mathbf{x}_{t+1}|\mathbf{x}_t) = \mathcal{N}(\mathbf{0}, \sigma^2 I) \\ p(\mathbf{y}_t|\mathbf{x}_t) \quad = \mathcal{N}(\boldsymbol{\mu}_{obs}, \sigma_{obs}^2) \end{cases} \tag{10}$$

Dynamical Bayesian networks describe relations of conditional dependencies on time-dependent variables. The proposed model also belongs to the family of *Kalman filter models* which was proposed in the late sixties by Kalman to process signal filtering and smoothing. In the following, we will refer to our model as the Inertial Dynamic Bayesian Network (IDBN).

4.3 The optimization scheme

Definition of a cost function

In the context of learning from a unique time-serie[5], we can define a cost function that should be minimized to get good generalization abilities. Within the probabilistic

[5] Note that the whole approach can be easily generalized to multiple time-series

Fig. 5. The Inertial Dynamic Bayesian network developed in time

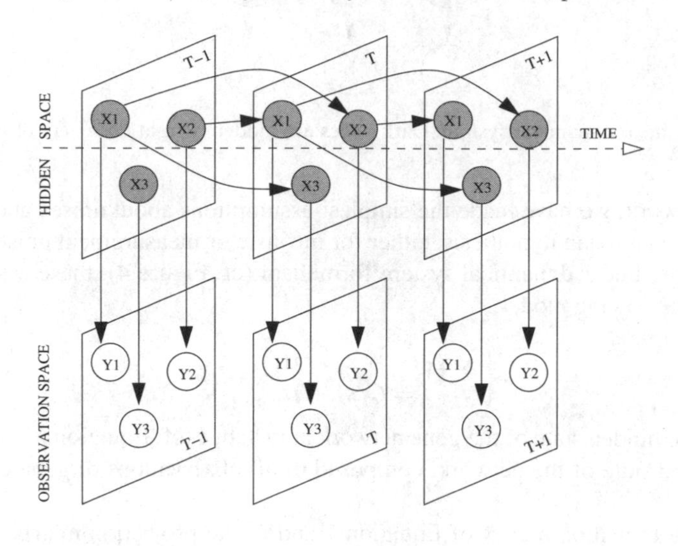

framework, a natural criterion is the log-likelihood L of the observed data denoted by $\mathbf{y}_{1:T}$ in our settings:

$$L(\theta; \mathbf{y}_1, ..., \mathbf{y}_T) = log\ p(\mathbf{y}_{1:T}|\theta) \tag{11}$$

In order to get a criterion over all the possible time-series, we can now derive from this definition the functional risk $R(\theta)$ as the expectation of minus the log-likelihood calculated over all time-series $\mathbf{y}_{1:T} = (\mathbf{y}_1, ..., \mathbf{y}_T)$:

$$R(\theta) = -\int L(\theta; \mathbf{y}_1, ..., \mathbf{y}_T) p_{true}(\mathbf{y}_{1:T}) d\mathbf{y}_{1:T} \tag{12}$$

However, as the true density p_{true} is not known (it is what we are looking for), this quantity cannot be calculated and another criterion needs to be defined. Both statistical learning [25] and regularization theories [8] provide us fundamental results that prone to replace this criterion by the combination of two criteria : one is based on the empirical estimation of $R(\theta)$, the other controls the complexity of the model. The underlying idea beyond this is that as far as the model fits correctly the data while it keeps simple, there will be no overfitting. Therefore, the obtained model will be enough general to deal with other time-series generated from the same distribution.

The empirical cost function will be here the **log-likelihood** of the observed data given in equation 11. Defining a criterion that reflects the complexity of the model appears to be a more difficult point.

However if we consider the underlying graph of interactions between genes which is represented by the W matrix, we can assume that a simple model for genetic interactions is a sparse model with few connections between genes. From a biological point of view this assumption corresponds to the observed fact that each gene

Filter
$\mathbf{x}^{t-1}(t) = A\mathbf{x}^{t-1}(t-1)$
$\Sigma^{t-1}(t) = A\Sigma^{t-1}(t-1)A' + \sigma_s^2 I$
$\Sigma_e(t) = C\Sigma^{t-1}(t)C' + \sigma_{obs}^2 I$
$K_t = \Sigma^{t-1}(t)C'\Sigma_e^{-1}(t)$
$\mathbf{e}_t = \mathbf{y}_t - C\mathbf{x}^{t-1}(t) - \boldsymbol{\mu}_{obs}$
$\mathbf{x}^t(t) = \mathbf{x}^{t-1}(t) + K_t\mathbf{e}_t$
$\Sigma^t(t) = \Sigma^{t-1}(t) - K_t C \Sigma^{t-1}(t)$

Table 1. Filter equations. $\mathbf{x}^0(1) = \boldsymbol{\mu}_i$, $\Sigma_1^0 = \sigma_i^2 I$, $\hat{\mathbf{x}}(T) = \mathbf{x}^T(T)$ and $\hat{\Sigma}(T) = \Sigma^T(T)$. parameters.

interact with very few other genes[6] In the optimization scheme,controlling the complexity of the model means to encourage sparsity of the W matrix during learning and this can be operated by minimizing the norm $||W||$ of matrix W.

Another equivalent way to interpret this combination of criteria is to consider the parcimony assumption as a prior in the framework of bayesian inference. Instead of minimizing only $-logP(\mathbf{y}_{1:T}|\theta)$, we will minimize the following quantity $C(\theta; \mathbf{y}_1, ..., \mathbf{y}_T)$:

$$C(\theta; \mathbf{y}_1, ..., \mathbf{y}_T) = -log\ p(\mathbf{y}_{1:T}|W, \theta') - log\ p(W) \tag{13}$$

with $p(W) \propto exp(-\alpha \sum_{ij} |w_{ij}|)$,assuming that the prior probability on W is Gaussian using L_1 norm.α is called the regularization hyperparameter: it weights the effect of the parcimony constraint applied to the parameters.

The optimization algorithm

Parameters can be learned using a generalized *Expectation-Maximization*(EM) algorithm [4, 1] and which has been thoroughly described for linear Gaussian processes in [20] in a unifying review.

EM algorithm, which allows to handle the hidden variables $\mathbf{x}_t$, is an iterative process that uses the following two steps :

The standard *Expectation* phase implements the *filter* and *smoother* processes, as described in [19] and summarized in Table 1 and Table 2. It allows to determine directly the most probable states $\mathbf{x}_t$ given $\mathbf{y}_{1:T}$.

The *Maximization* phase is different from the usual one, because of the specific nature of matrix A.

The auxiliary function $Q(\boldsymbol{\theta}, \boldsymbol{\theta}^{(k)})$, parameterized by $\boldsymbol{\theta}^{(k)}$, the current estimation of parameters θ at step k, is defined as the expectation (operator $E[\cdot]$) of the penalized log-likelihood [17] with respect to $\mathbf{y}_{1:T}$:

[6] Statistics of input and output degrees for instance for the yeast have been studied in [9].

Smoother
$$\begin{aligned} J_{t-1} &= \Sigma^{t-1}(t-1)A'(\Sigma^{t-1}(t-1))^{-1} \\ \hat{\mathbf{x}}(t-1) &= \mathbf{x}^{t-1}(t-1) + J_{t-1}(\hat{\mathbf{x}}(t) - \mathbf{x}^{t-1}(t)) \\ \hat{\Sigma}(t-1) &= \Sigma^{t-1}(t-1) + J_{t-1}(\hat{\Sigma}(t) - \Sigma^{t-1}(t))J'_{t-1} \\ \hat{\Sigma}^{t-1}(t) &= \hat{\Sigma}(t)J'_{t-1} \end{aligned}$$
$$\begin{aligned} \hat{R}(t) &= \hat{\Sigma}(t) + \hat{\mathbf{x}}(t)\hat{\mathbf{x}}'(t) \\ \hat{R}^{t-1}(t) &= \Sigma^{t-1}(t) + \hat{\mathbf{x}}(t)\hat{\mathbf{x}}'(t-1) \end{aligned}$$

Table 2. Smoother equations. $\hat{R}(t)$ and $\hat{R}^{t-1}(t)$ are used for determining the model parameters.

$Q(\boldsymbol{\theta}, \boldsymbol{\theta}^{(0)}) = E[\mathcal{L}^{pen}(\boldsymbol{\theta})|\mathbf{y}_{1:T}, \boldsymbol{\theta}^{(k)}]$.

The M phase used for our algorithm determines $\boldsymbol{\theta}^{(k+1)}$ by making a gradient step in the direction $\nabla_{\boldsymbol{\theta}} Q(\boldsymbol{\theta}, \boldsymbol{\theta}^{(k)})$ from $\boldsymbol{\theta}^{(k)}$, parameters estimated after k EM iterations:

$$\boldsymbol{\theta}^{(k+1)} = \boldsymbol{\theta}^{(k)} + \eta \nabla_{\boldsymbol{\theta}} Q(\boldsymbol{\theta}, \boldsymbol{\theta}^{(k)}) \quad , \quad \eta > 0 \tag{14}$$

More precisely, the M phase consists in the following computations:

- $\boldsymbol{\mu}_i^{(k+1)} = (1-\eta)\boldsymbol{\mu}_i^{(k)} + \eta \hat{\mathbf{x}}(1)$
- $\sigma_i^{2(k+1)} = (1-\eta)\sigma_i^{2(k)} + \eta \left[\frac{1}{2n}\hat{R}(1) - \boldsymbol{\mu}_i^{(k)}\boldsymbol{\mu}_i^{(k)'}\right]$
- $\boldsymbol{\mu}_{obs}^{(k+1)} = (1-\eta)\boldsymbol{\mu}_{obs}^{(k)} + \eta \left[\frac{1}{T}\sum_{t=1}^{T}(\mathbf{y}_t - C\mathbf{x}_t)\right]$
- $\sigma_{obs}^{2(k+1)} = (1-\eta)\sigma_{obs}^{2(k)} +$
  $\eta \left[\frac{1}{nT}\left(\sum_{t=1}^{T}(\mathbf{y}_t\mathbf{y}'_t - C\hat{\mathbf{x}}(t)\mathbf{y}'_t - \boldsymbol{\mu}_{obs}^{(k)}\mathbf{y}'_t)\right)\right]$
- $\sigma_x^{2(k+1)} = (1-\eta)\sigma_x^{2(k)} +$
  $\eta \left[\frac{1}{n(T-1)}\left(\sum_{t=2}^{T}(\hat{R}(t) - A^{(k)}\hat{R}'^{t-1}(t))\right)\right]$

For updating parameters $W^{(k)}$, $\Omega^{(k)}$ and $\Lambda^{(k)}$, it is useful to consider the G_k matrix defined by:

$$G_k = \frac{1}{\sigma_x^{2(k)}} \sum_{t=2}^{T} \left[\hat{R}^{t-1}(t) - A^{(k)}\hat{R}(t-1)\right] \tag{15}$$

and more precisely the matrices of size $n \times n$ $G_k^{\ell\ell}$ and $G_k^{\ell r}$ which denote respectively the G_k lower left and lower right submatrices. Parameters $W^{(k+1)}$, $\Omega^{(k+1)}$ and $\Lambda^{(k+1)}$ are computed according to:

$$\begin{aligned} W^{(k+1)} &= W^{(k)} + \eta(G_k^{\ell\ell} - \lambda \nabla_W ||W||) \\ \Omega^{(k+1)} &= \Omega^{(k)} - 2\eta \operatorname{diag}(G_k^{\ell\ell} \Omega^{(k)} + G_k^{\ell r} \Lambda^{(k)}) \\ \Lambda^{(k+1)} &= \Lambda^{(k)} - \eta \operatorname{diag}(G_k^{\ell r} \Omega^{(k)}) \end{aligned}$$

Once the $A^{(k+1)}$ matrix is computed according to Equation (7), it is possible to proceed to a new E phase. At each EM step, the penalized likelihood increases, until a local maximum is reached.

Handling missing variables

Let us notice an important feature of the IDBN model : if some genes are to play a decisive role in the network, but if no expression data is available for them, it is however possible to include them by modifying some of the model features. These genes will be called *missing variables*. The proposed framework is adapted to handle them: the model is divided in two different spaces, the *hidden space* of $\mathbf{x}_t$ and the *observation space* of $\mathbf{y}_t$; there is only a projection matrix C to go from hidden state to observed state. Let us study a network composed of $n+h$ units, with available data for n genes, but the others h genes being unmeasured. Vectors $\mathbf{y}_t$ of the observation space do not change and are always of size n. On the contrary, vectors $\mathbf{x}_t$ are now of size $2(n+h)$, being composed of all genes, including the missing ones. A is now a $2(n+h) \times 2(n+h)$ matrix, and the projection matrix C is $[I \quad \mathbf{0_{n,h}} \quad \mathbf{0_{n,n+h}}]$ of size $n \times 2(n+h)$. Incorporating missing variables in our framework is therefore very easy, and the EM learning algorithm will learn them as unknown parameters.

5 Experiments

5.1 Experimental data sets

In order to validate our approach, we considered the *S.O.S. DNA Repair* network of the *Escherichia coli* bacterium. The goal of is to learn the parameters of the inertial model in order to fit the available experimental data. No prior knowledge on the structure of the network is used except the average connectivity degree which is supposed to be given. Validation is first realized by measuring how the identified model is able to do prediction one or k step ahead and by comparing betwen the learned model and the true network.

The well-known regulation network, *S.O.S. DNA Repair*, is responsible for repairing the DNA after a damage. The whole system is composed of about 30 genes regulated at the transcriptional level. Usually, when no DNA damage occurs, a master transcription factor *LexA* binds sites in the promoter regions of these genes, repressing all genes of the network. One of the *S.O.S.* proteins, *RecA*, acts as a sensor of DNA damage: by binding to single-stranded DNA, it becomes activated and mediates *LexA* destruction. The drop in *LexA* levels causes the de-repression (i.e. activation) of *S.O.S.* genes. Once damage has been repaired or bypassed, the level of activated

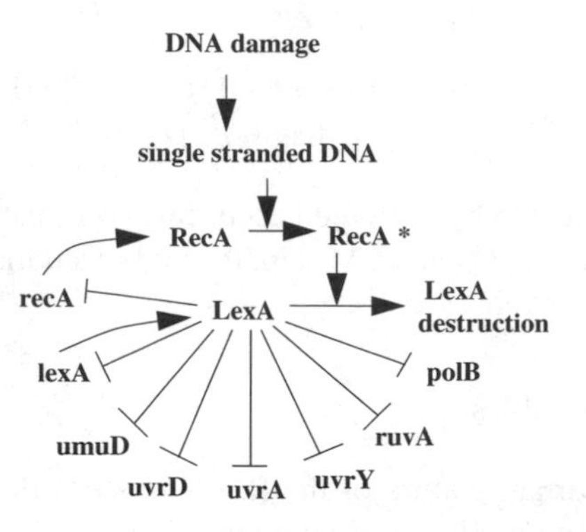

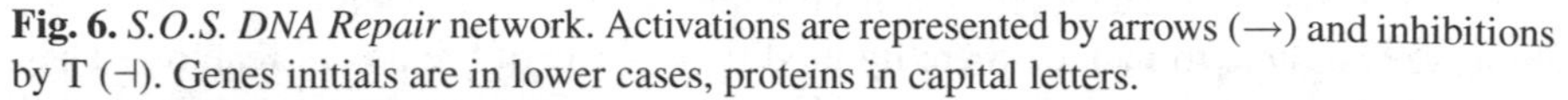

Fig. 6. *S.O.S. DNA Repair* network. Activations are represented by arrows (→) and inhibitions by T (⊣). Genes initials are in lower cases, proteins in capital letters.

RecA drops, *LexA* accumulates and represses the *S.O.S.* genes, and cells return to their initial state.

Experimental data have been made available by Uri Alon (they are downloadable on its homepage [7]). Data are expression kinetics of the main 8 genes of the *S.O.S. DNA Repair* network of *E.coli*. The measurement technology is based on the property of the GFPs (green fluorescent proteins). Alon et al. have developed a system for obtaining very precise kinetics [18]. Measurements are done after irradiation of the cells at the initial time with UV light. Four experiments are done for various light intensities (Exp. 1&2 : 5 Jm^{-2}, Exp. 3&4 : 20 Jm^{-2}). Each experiment is composed of 50 instants evenly spaced by 6 minutes intervals, and 8 genes are monitored: *uvrD*, *lexA*, *umuD*, *recA*, *uvrA*, *uvrY*, *ruvA* and *polB*.

We first explain why the data are useful for our purpose and can be incorporated in our model.

Alon et al. monitor each *S.O.S.* gene separately, adding its promoter to a gfp sequence in a plasmid and incorporating the plasmids in irradiated *E.coli*. The quantity of present GFP, which is indirectly measured by the amount of fluorescence, is therefore proportional to the quantity of the corresponding *S.O.S.* protein.

The first hypothesis made by Alon et al. is that the GFP protein is very stable during the experiments: it is justified when comparing the typical GFP stability to the experiments length (300 minutes). By measuring the time-course of fluorescence intensity, Alon and coworkers have therefore access to the instantaneous protein production rate, since no protein degradation occurs during the experiments.

By making the standard hypothesis that the protein production rate is proportional to the corresponding mRNA production rate, they consider that the derivatives of the fluorescence amounts are proportional to the promoter activity of the genes. It is important to notice that this hypothesis is not so hard as usual in this case, because all proteins are the same (GFP protein). One more difficult point is why Alon et al.

[7] http://www.weizmann.ac.il/mcb/UriAlon/

divide also by the OD, but we will not discuss this point here. Figure 3 (b) of [18] indicates these promoter activities.

To use these data in our model, we have to make a strong hypothesis on the stability of the mRNAs: we consider that mRNA molecules are degraded immediately after their production. Actually mRNA persistance depends on the nucleotide sequence; all mRNAs having the same sequence here because of the experimental technique, they all have the same persistance. It is hence sufficient to make the hypothesis that the turnover of the GFP mRNAs is very fast to ensure that there is a high unstability of the mRNAs. We therefore consider that the instantaneous promoter activity of each gene is also proportional to the present quantity of corresponding mRNA.

We hence are able to consider that the data provided by Alon et al. directly indicate the observed mRNA quantities (also called expression levels) corresponding to each *S.O.S.* gene. The downloaded data consists in four 8×50 matrices corresponding to the four stresses. The tth column of a given matrix is considered to be the observation vector $\mathbf{y}_t$, and the entire matrix will be considered as the time course $\{\mathbf{y}_1, \ldots, \mathbf{y}_{50}\}$.

5.2 Learning experiments

We have proceeded to several learning experiments on the data provided by Alon et al. with the Inertial Dynamical Bayesian Network (IDBN). The influence of the regularization parameter α is important and has been carefully studied. For each data set, we have made successive learnings with α =0, 1, 5, 10, 50, 100, 500, 1000 and 5000. In each case, we have introduced 0 or 1 missing variable. The EM algorithm stops after 100 iterations, which is sufficient, since not discussed here previous experiments have shown that after 80 iterations, more than 95% of the parameters will not change of more than 1%, which does not change significantly the learned model. Parameters are initialized as follows: each coefficient of W is randomly initialized between 0.01 and 0.01, Λ and Ω are chosen from the assumption that unoscillatory behaviour is observed ($\lambda_i \simeq 1$ and $\omega_i \simeq 0\ \forall i$), each coefficient of μ_i and μ_{obs} between 0 and 1, and finally σ_i^2, σ_x^2 and σ_{obs}^2 between 0 and 0.1. The main crucial point seems to be the initialization of W, Λ and Ω. 50 different learnings under each condition have hence been made: this *multiple random starting points* technique is widely used to handle the problem of likelihood local maxima.

5.3 Identifying the network dynamics

The proposed method is able to capture the network dynamics. Figure 7 allows to compare the real profiles of the 8 genes relative to the second experiment and their simulated profiles corresponding to the learned model for $\lambda = 100$. It is important to notice that these simulated profiles are *mean profiles*, since the variances associated to the model (σ_i^2, σ_x^2, σ_{obs}^2) are not taken into account in this simulation. The behaviour of the network according to these elements will be discussed later. The variations of the most expressed genes of the network (*recA*, *lexA*, *uvrA*, and to a lesser extent *umuD* and *uvrD*) are finely modelled. One can notice that the respective maxima of these genes are reached at different instants whose succession is essential to explain the functioning of the *S.O.S.* response system: this succession is respected in the learned model.

No noteworthy dose effect has been observed when comparing results of the different experimental conditions. It only appears that gene expressions levels are higher when the irradiation intensity increases, but the respective maxima and variations do not change between the four experiments, indicating that the *S.O.S.* system has the same type of answer for both experiment intensities.

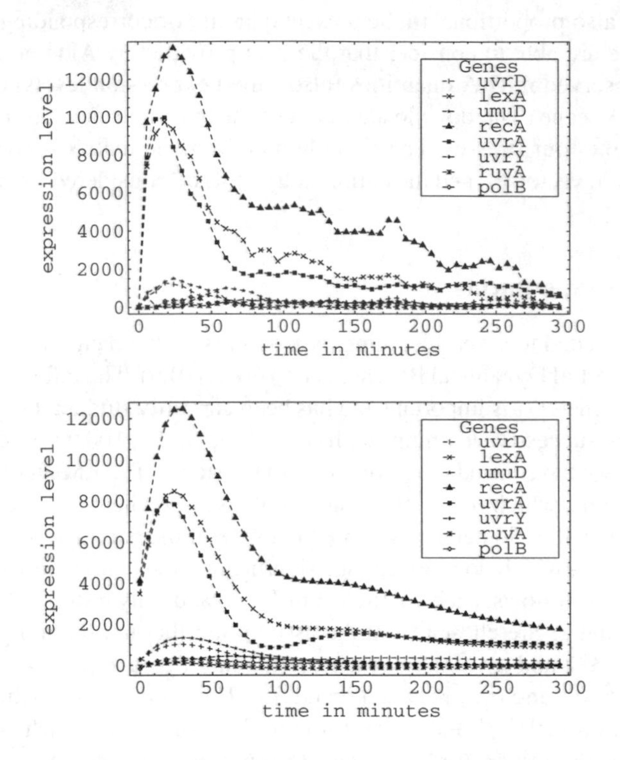

Fig. 7. Top: expression profiles of the 8 genes corresponding to the second second data set. Bottom: learned profiles from this data set for $\lambda = 100$. The vertical scale has no absolute meaning, since we took the (dGFP/dt)/OD values of [18], because they are proportional to the expression levels of the genes as explained previously. It is important to notice that these simulated profiles are *mean profiles*, since the variances associated to the model are not taken into account in this simulation.

Extracting the stochastic phenomena

One of the characteristics of our model is its ability to represent stochastic phenomena: one could wonder if the stochastic learned parameters are compatible with the real data. On Figure 8, we have represented five simulated profiles of *lexA* and *recA* using the model learned from the second data set with $\alpha = 50$.

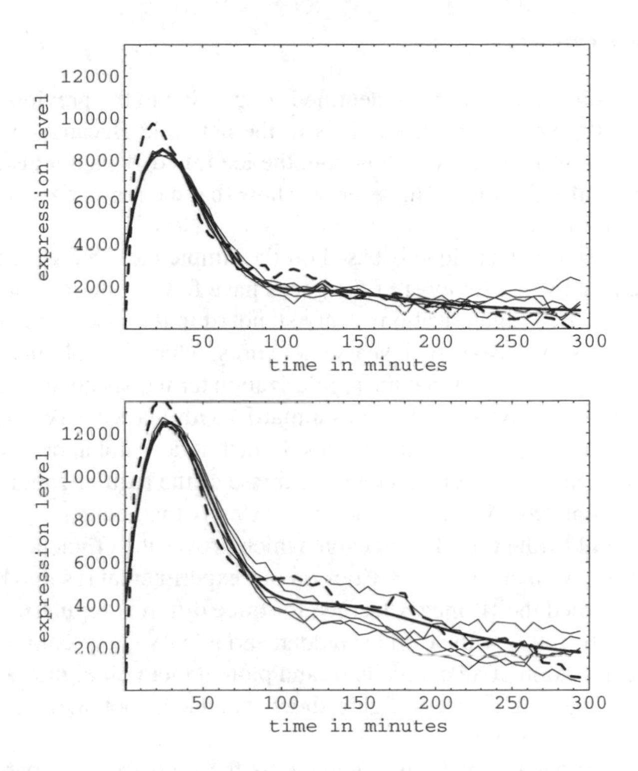

Fig. 8. Profile of *lexA* (top) and *recA* (bottom) under the same conditions as in Figure7. The thick dotted line is the measured profile given in the data set; the thick continuous line is the mean learned profile already represented on Figure 7; thin continuous lines are simulated profiles, taking into account the learned model variances.

For simulating such profiles, we use Equations 9. At each time point, $\mathbf{u}$ and $\mathbf{v}$ are randomly choosen according to their Gaussian distribution with variances σ_x^2 and σ_{obs}^2. The system is also randomly initialized according to $\mathbf{x}_1 \sim \mathcal{N}(\mu_i, \sigma_i^2 I)$. We also have represented the mean profiles as in Figure 7 setting the variances to 0, and the measured profiles of the second data set. One can notice that the measured profile and the mean simulated profile do not merge: nevertheless, the measured profile is located in the enveloppe of simulated stochastic profiles (at least during the decrease). This shows that the learned variances are compatible with real variances.

For lach of space, we do not have plotted here such simulated profiles for all the genes under others conditions. Nevertheless experiments allow us to conclude that model variances associated with other learned model parameters are in accordance with the observed profiles. Our learning technique associated with the model choice is hence able to capture the mean dynamics of the network, so as to take into account stochastic phenomena, either due to measurement noise or to regulation itself.

5.4 Structure extraction

After each learning, a W matrix is identified. According to the previous assumption, it should directly indicate the regulations in the network. Because of the presence of local maxima of the likelihood function, the identified W is not always the same at each numerical experiment. However we show that the regularization process can alleviate the problem.

The regularization technique is based on the simple idea that gene networks are known to be usually sparse: most of the genes have few regulators, and in turn regulate few genes. This regularization is very standard in the machine learning framework and is known to favour such sparse networks, which is biologically motivated in our case. One could object that the regularization term does not encourage sparseness, but simply low norm for W, so as a matrix with many weak connections can be as favorable as one with few strong ones. In fact, it does not appear to behave like that. Of course, each w_{ij} coefficient will decrease as the regularization parameter α increases, but some coefficients decrease more slowly than others.

In order to illustrate this phenomenon, which proves the efficiency of our penalizing term, we have made some statistics on our experimental results. For each data set, we have studied the W matrix learned for three different regularization parameters $\alpha = \{0, 100, 1000\}$ with no added hidden variable. We have computed the mean and standard deviation of all coefficients, and plotted them on Figure 9. Coefficients are ordered by mean. The plotted figure shows the results obtained for the first data set, but the others are similar.

It seems obvious that the curve flattens as the regularization parameter α increases. This results proves that our regularization term encourages a real sparseness, and not only low coefficients. Of course, one can notice that all means decrease with α, but some coefficients remain strictly non zero: this is the main goal of our regularization. Moreover, we also can notice that standard deviations decrease with α: it should suggest us that a good regularization decreases the search space dimension, and in consequence the number of local maxima of the likelihood function. Nevertheless, when α is too high, all coefficients tend to zero, so that all interesting information about regulations vanish.

There are several ways to select or learn α value. If numerous independent time series were available, it would be possible to apply cross-validation to select its right value. Another method consists in applying a full Bayesian approach by assuming a prior on α and incorporating in the algorithm a learning stage for α. In our framework, the regularization parameter can be chosen according to the average degree d of the genetic graphs. The average degree of a graph is given by $d = \frac{2M}{N}$ where N is the number of nodes, and M the number of arcs. It indicates the average number of arcs bound to a node. Some data are available about the topological structure of transcriptional networks: [22] give 577 interactions for 116 transcription factors in *E.coli* using the RegulonDB database, which leads to $d = 9.95$, while [9] propose

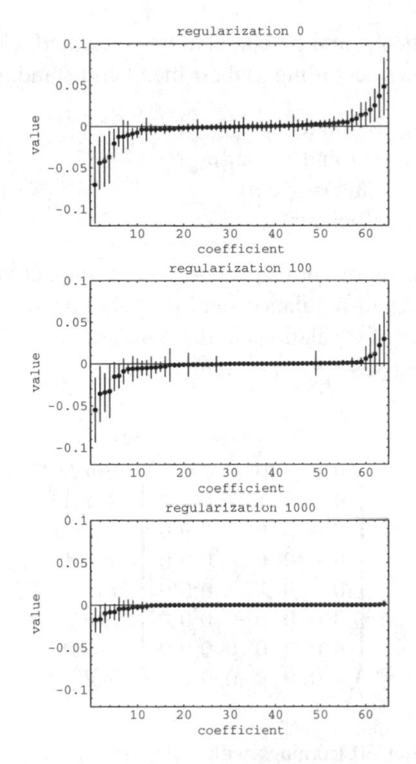

Fig. 9. W coefficients learned from the first data set, using respectively regularization 0, 100 and 1000. Coefficients α are ordered by their mean on the 50 learnings. Error bars indicate their standard deviation.

a yeast transcriptional network composed of 491 genes and 909 transcriptional interactions giving $d = 3.70$. Unfortunately, the *S.O.S.* network has a very particular *star-like* topology, and is much smaller. We hence cannot use these average degrees for finding the best regularization. For this experiment, we hence use the average degree of the actual network. Figure 6 gives $d = 2.25$ for transcriptional regulations in the *S.O.S.* network considering that there are 9 transcriptional regulations between the 8 genes. As we find $d = 4.25$ for $\alpha = 0$, $d = 2.25$ for $\alpha = 100$, and $d = 1$ for $\alpha = 1000$ (the way to obtain the number of identified regulations is discussed further), we will consider that the optimal regularization parameter is 100. This regularization allows to make a compromise between favouring the sparseness of W and keeping information about regulations.

Identifying the structure of the network

For each data set, with $\alpha = 100$, 50 learnings starting from random starting points are done as explained previously. The learned values of each parameter w_{ij} are distributed with mean μ_{ij} and variance σ_{ij}^2. The mean and variance of the means of all 64 coefficients named μ and σ^2 can also be computed. Coefficients are then discretized into four classes according to their mean and standard deviation:

- class `[+]` : $\mu_{ij} > \mu + \sigma$ and $\sigma_{ij} < |\mu_{ij}|$
- class `[-]` : $\mu_{ij} < \mu - \sigma$ and $\sigma_{ij} < |\mu_{ij}|$
- class `[0]` : $|\mu_{ij}| < \sigma$ and $\sigma_{ij} < \sigma$
- class `[X]` : others coefficients

Classes are built to represent respectively probable activations, probable inhibitions, probable absences of regulation, and probable presences of unknown regulations. The total number of regulations in the network is obtained adding the number of coefficients in the classes `[+]`, `[-]` and `[X]`.

$$\begin{pmatrix} 0&0&0&0&0&0&0&0 \\ 0&-&+&+&-&0&0&0 \\ 0&0&0&0&0&0&0&0 \\ 0&-&0&0&-&0&0&0 \\ 0&-&0&X&X&0&0&0 \\ 0&0&0&0&0&0&0&0 \\ 0&0&0&0&0&0&0&0 \\ 0&0&0&0&0&0&0&0 \end{pmatrix} \begin{matrix} uvrD \\ lexA \\ umuD \\ recA \\ uvrA \\ uvrY \\ ruvA \\ polB \end{matrix}$$

Fig. 10. W identified after 50 learnings with $\alpha = 100$ on the first data set. Discretization process is described in the article. The jth column shows all identified regulations exerced by j^{th} gene on other genes. Inversely, the ith row shows all regulations the ith gene is submitted to. Genes order is on the right.

Figure 10 shows the identified discretized W matrix using the first data set. One can notice that 9 probable regulations are identified in the network, which leads to an average degree of 2.25, as said previously. Inhibitions of *lexA*, *recA* and *uvrA* by *lexA* itself are well identified (second column). The activation of *lexA* by *recA* is also identified (W_{24}). False regulations due to *umuD* and *uvrA* (columns 3 and 5) are identified (these genes are target genes, and do not act as regulators). A unknown regulation of *uvrA* by *recA* is identified: it could correspond to the indirect regulation *recA* $\rightarrow$ *RecA* $\dashv$ *LexA* $\dashv$ *uvrA*.

For further comparisons between an identified network and the actual network, we have to specify the perfect W matrix which should correspond to the *S.O.S.* network. Of course, as some regulations are not directly transcriptional (for example the activation of *uvrA* by *recA* seen previously) the choice is not immediate. We shall consider that all parameters of the second column, indicating inhibitions on all genes by *lexA* have to be discretized in the class `[-]`, while the activation of *lexA* by *recA* is represented by W_{24} in `[+]`. Other parameters of the fourth columns are not defined precisely and can be either in `[0]` or `[+]`, because the learning is likely to identify undirect regulations.

Structure extraction can also be viewed as an information retrieval task. We can transpose *Recall* and *Precision* measures usually used as quality measures in document retrieval to our field of interest. The same idea has been developed by [24] and more cently bu [11]. *Recall* defines the number of true oriented interactions predicted as fraction of all existing interactions. *Precision* defines the number of true oriented interactions predicted as fraction all the interactions predicted. Precision thus defines the level of "noise" in the information presented to the user.

But, it should be emphasized that structure extraction not only aims at finding regulations, but also at identifying absence of interactions. We should hence also measure the number of true predicted coefficients as fraction of all interactions between genes (regulations and non regulations). We define this number as the *Generalized precision*.

Recall mean on all 4 data sets is 0.66, with a standard deviation of 0.056; *Precision* gives a mean of 0.57, with a standard deviation of 0.083; *Generalized precision* is 0.87 associated with a standard deviation of 0.023. For a random matrix, one should notice that *Recall* = 0.60, *Precision* = 0.20, and *Generalized precision* = 0.31. *Generalization* and *Generalized performance* are significantly high. *Recall* seems to be quite low, but we have to remember that the true "score" of our method is given by the *Generalized precision*, as biologists are not only interested in regulations, but also in the absence of regulations. However the value of the *Recall* needs a comment: it can be enlightened by the fact that the experimental UV light shock was not sufficient to lead to the functioning of all *S.O.S.* genes. Figure 7 (top) shows that several genes were not induced during the experiment. These genes are activated only when the damage is sufficiently high.

In order to evaluate the similarity between networks identified using the various data sets, the proportion of equal coefficients between these networks are computed, giving a similarity mean of 89 % with a standard deviation of 11 %. It should be emphasized that these similarities are high, comparing with 25 % obtained with a random matrice. These high similarities show that several experiments on the same underlying network let similar networks be identified.

5.5 Dealing with a missing variable

The *S.O.S.* DNA repair of *E. coli* considered here involve both genes and some of their products, proteins. The fact that neither the level of *LexA* nor the level of *RecA* protein are available makes it impossible to retrieve the true direct regulation pathways. It is especially worrying in the case of LexA which plays a central role. That is the reason why we decided to use one important feature of both the inertial model and the EM algorithm, the handling of a missing variable.

When one missing variable is introduced, simulations of its level evolution are done using each of the 50 learned models. For 22 of these models, it is quite noticeable that the simulated profile is akin to the *LexA* protein concentration profile under the same experimental conditions measured in [21]. We present an example of simulated profiles using one of these 22 solutions in figure 11.

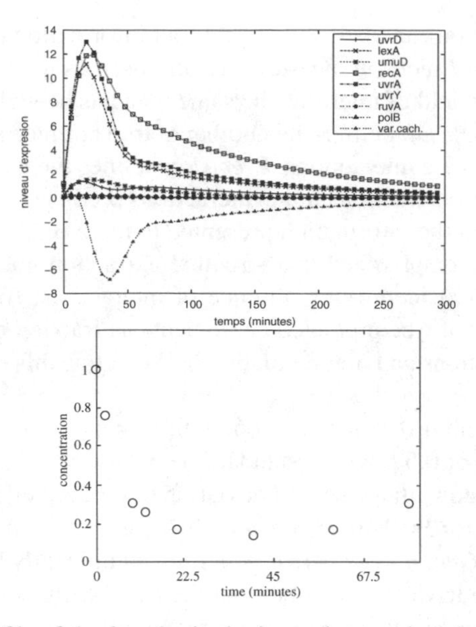

Fig. 11. Simulated profile of the 9 nodes in the learned network and profile of LexA protein such as the one measured by Sassanfar in [21] in similar conditions

Moreover, when considering only this cluster of models, variances concerning the regulations involving the added missing variable are lower than previously, so that W discretization shows that missing variable is inibited by *recA* and inhibits *lexA* and itself.

An example of obtained network is presented in the figure 12. A pruning algorithm that deletes redundant indirect interactions has been applied for clarity purpose.

An attractive hypothesis is that the added missing variable "takes the role" of the protein *LexA*. However further an dcomplete experiments need to be done to show if this hypothesis is acceptable.

5.6 Prediction

Within the machine learning theory, the choice of a model and the identification of its parameters should lead to generalization ability. For sequential data, this ability can be measured by two properties: the model ability to make k-step ahead prediction and its ability to reflect dynamics of other i.i.d. sequences. In the context of the available data, our learned model easily succeeded in making k-step prediction using the first 2/3 data points for training and the 1/3 lasting time for prediction. This does not prove very much since prediction is quite easy (back to equilibrium) but is needed to be checked.

We also used one time course as training data and others as test data. Data provided by Alon et al. are particularly adapted to this type of experiments, because the four available time-series concern the same network. Two kinds of prediction abili-

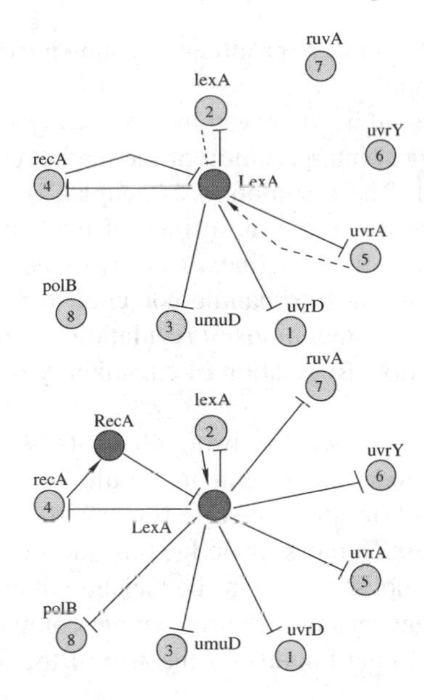

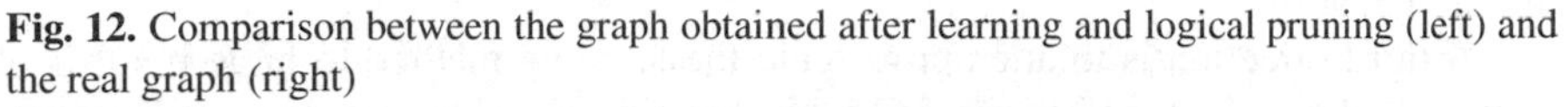
Fig. 12. Comparison between the graph obtained after learning and logical pruning (left) and the real graph (right)

ties are evaluated. The first one is called the *one time step prediction*: for each instant, an expectation of the observation at the next instant is computed using the test data observation and the model learned from the training data. The second prediction ability is called the *multi step prediction*: a filtering phase is done on the 10 first instants of the testing data to estimate its hidden state at instant 10, and the model learned from the training data is then used to predict the kinetic profiles of the genes until instant 50 without accessing to their true observation values.

Very good correlations between *one step predicted* sequences and actual sequences are achieved, with a mean of 0.968, and a standard deviation of $15.6\ 10^{-3}$. Correlation between the last part (instants 11 to 50) of *multi step predicted* sequences and actual sequences has a mean of 0.654 and a standard deviation of 0.171. Even if the predicted part of the sequence is the easiest because of the monotonous decrease of gene profiles after their peak, such correlations prove that the learned models are able to predict further evolution.

6 Conclusion and Future Work

We presented a general methodology for reverse modeling based on dynamical modeling and machine learning tools. We instantiated this approach to a new model of dynamics that shares the advantages of additive models while keeping into account delays and inertia in the response of a regulated gene. The first results on the *S.O.S.*

DNA Repair network of *E.coli* were promising by showing the power of our approach and of the considered model.

However this work could be improved in several ways. First, from the modeling point of view, the biological interpretation of the mathematical model we proposed can be further discussed. The assumption of coupled oscillatory behaviours even dampened for genes is at this stage an artifice of mathematicians. It allows us to capture complex behaviours within a linear framework but can be challenged within a non-linear context for instance to handle conjunctive regulations. A future work will also consist on studying other kinds of regulations where more belief can be put in the oscillatory assumption. Regulation of circadian rythms for instance can be an example of such regulations.

As a second direction, we want to bring elements of answer to a crucial point for the biologists : how many measurements should be produced to make possible the identification process? In our time-serie framework, it reduces to the length of the avalaible time-courses. To our knowledge this question has not yet been solved even for linear dynamic models. It has to be taclked within the theory of statistical learning[25] and its recent results that link sample complexity to the model complexity. The principle is to get bounds on the size of the data sets that allow good prediction ability.

A third direction is to attempt to scale the learning method to large number of genes. At this stage for large networks, our learning algorithm meets too many local minima. Applying a "divide and conquer principle" is a possible way to solve this question. Some authors have already used this approach to reduce large networks to a set of connected subnetworks for each of which learning is easier (see [14],[3] for instance). Another way to address the question consists in taking into account more biological knowledge. Gene annotation can provide at this stage many informations such as gene functions and binding sites. A full bayesian approach applied to the dynamical bayesian network could be of help for incorporating such prior knowledge.

References

1. J. Blimes. A gentle tutorial of the em algorithm and its application to parameter estimation for gaussian mixture and hidden markov models, 1998.
2. T. Chen, Hongyu L. He, and George M. Church. Modeling gene expression with differential equations. In *PSB*, 1999.
3. Michiel J. L. de Hoon, Seiya Imoto, and Satoru Miyano. Inferring gene regulatory networks from time-ordered gene expression data using differential equations. In *Proceedings of the 5th International Conference on Discovery Science*, pages 267–274. Springer-Verlag, 2002.
4. P. A. Dempster, M. N. Laird, and B. D. Rubin. Maximum likelihood from incomplete data via the EM algorithm. *JRSSB*, 39(1-38), 1977.
5. P. D'Haeseleer, S. Liang, and R. Somogyi. Genetic network inference: From co-expression clustering to reverse engineering. *BioInformatics*, 16(8):707–726, 2000.
6. N. Friedman, M. Linial, I. Nachman, and D. Pe'er. Using bayesian networks to analyze expression data. In *RECOMB*, pages 127–135, 2000.

7. Z. Ghahramani and G. E. Hinton. Variational learning for switching state-space models. *Neural Computation*, 12(4):831–864, 2000.
8. F. Girosi, M. Jones, and T. Poggio. Regularization Theory and Neural Networks Architectures. *Neural Computation*, 7(2):219–269, 1995.
9. N. Guclzim, S. Bottani, P. Bourgine, and F. Képès. Topological and causal structure of the yeast transcriptional regulatory network. *Nature Genetics*, pages 60–63, 2002.
10. D. Heckerman. A tutorial on learning with bayesian networks. Technical Report MSR-TR-95-06, Microsoft Research, 1995.
11. D. Husmeier. Sensitivity and specificity of inferring genetic regulatory interactions from microarray experiments with dynamic bayesian networks. *Bioinformatics*, 19(17):2271–2282, 2003.
12. S. Kim, S. Imoto, and S. Miyano. Dynamic bayesian network and nonparametric regression for nonlinear modeling of gene networks from time series gene expression data. In *Proc. of CMSB 2003*, pages 104–113, 2003.
13. H. McAdams and A. Arkin. Stochastic mechanisms in gene expression. *Proc. Nat. Acad. Sci.*, pages 814–819, 1997.
14. E. Mjolness, T. Mann, R. Castao, and B. Wold. From co-expression to coregulation: An approach to inferring transcriptional regulation among gene classes from large-scale expression data. In *Advances in Neural Information Processing Systems*, volume 12, pages 928–934. 2000.
15. K. Murphy and S. Mian. Modelling gene expression data using dynamic bayesian networks. Technical report, University of California, Berkeley, 1999.
16. I. Ong, J. Lasner, and D. Page. Modelling regulatory pathways in e.coli from time series expression profiles. *Bioinformatics*, 18:s241–s248, 2002.
17. D. Ormoneit and V. Tresp. Averaging, maximum penalized likelihood and bayesian estimation for improving gaussian mixture probability density estimates. *IEEE Transactions on Neural Networks*, 9, 1998.
18. M. Ronen, R. Rosenberg, B.I. Shraiman, and Uri ALon. Assigning numbers to the arrows:parameterizing a gene regulation network by using accurate expression kinetics. *PNAS*, 2002.
19. A.-V.I. Rosti and M.J.F. Gales. Generalised linear Gaussian models. Technical Report CUED/F-INFENG/TR.420, Cambridge University Engineering Department, 2001. Available via anonymous ftp from `ftp://svr-ftp.eng.cam.ac.uk/pub/reports/`.
20. S. Roweis and Z. Ghahramani. A unifying review of linear Gaussian models. *Neural Computation*, 11(2):305–345, 1997.
21. M. Sassanfar and J. Robert. Nature of the sos-inducing signal in escherichia coli. the involvement of dna replication. *J. Molecular Biology*, 212:79–76, 1990.
22. S. Shen-Orr, R. Milo, S. Mangan, and U. Alon. Network motifs in the transcriptional regulation network of escherichia coli. *Nature Genetics*, pages 64–68, 2002.
23. R. Thomas, D. Thieffry, and M. Kaufman. Dynamical behaviour of biological regulatory networks,i. biological role of feedback loops and practical use of the concept of the loop-characteristic state. *Bulletin of mathematical Biology*, (57), 1995.
24. E. van Someren, L. Wessels, and M. Reinders. Genetic network models: A comparative study, 2001.
25. V. Vapnik. *Statistical Learning Theory*. John Wiley and Sons, inc., 1998.
26. C. D. Weaver, T. C. Workman, and C. Storm. Modeling regulatory networks with weight matrices. In *Proc. of Pacific Symposium on Biocomputing*, volume 4, pages 112–123, 1999.

Class Prediction with Microarray Datasets

Simon Rogers[1], Richard D. Williams[2], and Colin Campbell[1]

[1] Advanced Computing Research Centre, University of Bristol, BS8 1TR, United Kingdom
C.Campbell@bristol.ac.uk
[2] Dept. of Paediatric Oncology, Institute of Cancer Research, Sutton, SM2 5NG, United Kingdom

Summary. Microarray technology is having a significant impact in the biological and medical sciences and class prediction will play an increasingly important role in the use and interpretation of microarray data. For example, classifiers could be constructed indicating the detailed subtype of a disease, its expected progression and the best treatment strategy. In this chapter we outline the main stages involved in the development of a successful class predictor for microarray datasets, including data normalisation, the different classifiers which can be used, different feature selection strategies and a method for determining how much data is required for a classification task given an initial sample set. We illustrate this process with both public domain datasets and a new dataset for predicting relapse versus non-relapse for a paediatric tumour.

1 Introduction

Microarray technology enables the simultaneous determination of the expression levels of thousands of genes. This has opened up a wealth of opportunities which could revolutionise our understanding in many areas of biological and medical research. cDNA and oligonucleotide microarrays have been used for a number of purposes. For example, in cancer research, comparisons of expression profiles have been used to find genes consistently over- or under-expressed in a tumour relative to a normal sample. In a comparison of samples drawn from multiple sclerosis lesions and normal tissue, microarrays highlighted a number of over-expressing genes associated with the immune response [9]. As an auto-immune disease this is to be expected. However, a further set of genes with unanticipated relevance was also found to be associated with the disease. In a study of schizophrenia, microarrays were used to compare expression profiles from different areas of the brain. For cells in the prefrontal cortex this highlighted the under-expression of genes associated with functioning of the pre-synaptic juntion [16].

Apart from discovering the significance of individual genes, collective expression profiles can give new insights. For example, cluster analysis has been used to detect previously unrecognised tumour subcategories. Alizadeh et al [6] analysed lymphoma tissue samples and found evidence for two previously unrecognised subtypes of this disease. These subtypes had distinct expression profiles at the molecular level and patients belonging to the two subclasses had different clinical prognoses. Similarly Bhattacharjee et al [7] and Sorlie et al [14] have identified subtypes of lung and breast cancer.

A further important application of microarray technology is class prediction. For cancer applications, classifiers could be built which may reliably indicate subtype, invasiveness potential, expected progression and the best treatment strategy. This could have important clinical consequences. For example, for prostate cancer, tumours can range from indolent to aggressive and microarray technology could be used to predict where on this range a new sample lies. If an indolent tumour type is reliably indicated, invasive treatment could be avoided and replaced by regular surveillence.

In this chapter we will discuss the main issues arising in the use of microarray data for class prediction. To build a successful classifier we first need to prepare and normalise the data. Performance of the classifier can be affected by the methods used for handling missing values or poor readings and by our use of normalisation within and between slides. Thus, in section 2 we will start by briefly discussing these issues with illustrations from real-life datasets. In section 3 we will discuss different types of classifiers which can be used with microarray datasets, the use of feature selection and methods which implement a confidence measure for the class assignment. For some classification tasks the available data may be insufficient in size to construct a classifier with good generalisation. Thus in section 4 we will consider the dependence of generalisation ability on sample size and how we may use the currently available data to determine how much data is required to achieve a given prediction performance. We will focus on cDNA and oligonucleotide arrays though related technologies such as small molecule or protein microarrays can pose similar class prediction problems.

2 Data preparation and normalisation

Successful data preparation obviously plays a crucial part in the construction of a reliable classifier since the experimental process can introduce serious artefacts in the data. The pre-processing steps required will depend on the experimental procedure used. For example, different procedures will be required for spotted cDNA and synthesised oligonucleotide microarrays (for example, for Affymetrix arrays most of the normalisations described below are commonly implemented by supplied data analysis software). In this section we will only sketch some of the main problems which can appear. We will assume that the ith slide in a microarray can be represented as a vector $\mathbf{x}_i$ with an associated class label y_i. The vector $\mathbf{x}_i$ has component *attributes* or *features* derived from the measured expression level of each spot on the microarray. Thus a feature may have a direct association with a particular gene or an expressed sequence tag (EST), or it may correspond to a replicate measurement or a probe with no known associations. Depending on the experimental procedure, it is common practice to first take the log of the data. Thus, for a cDNA array, suppose we derive expression ratios for a two channel experiment with one channel the control and the second derived from the sample of interest. For a doubling of expression over control the ratio is 2.0, whereas for a halving of expression the ratio is 0.5. The base 2 log of these ratios are 1.0 and -1.0 respectively, introducing a symmetry between over- and under- expression.

Next we need to normalise the dataset to adjust for any effects due to the experimental process rather than the biology. For cDNA microarrays, for example, there may be uneven hybridisation of dyes across the slide. This can give large scale systematic deviations in the data which negatively impact on eventual performance of the classifier. One approach is to determine a 2-dimensional Lowess (Loess) surface [26, 22] through the data to find any abnormal trends and correct the data values using this surface. The Lowess surface is a locally weighted polynomial regression surface with a low-degree polynomial fitted to a subset of the data around each point and the coefficients for this polynomial are found using a weighted least squares procedure which gives most weight to points nearest the point of interest. A global Lowess correction has the possible drawback that a localised cluster of differentially expressed genes could be biologically significant, but they could be modified by this spatial trend adjustment. Alternative strategies include step-wise print-tip and scaled print-tip normalisations [17].

Apart from uneven hybridisation of the dyes, the dyes themselves may have different dynamic ranges. In Figure 1 we show a plot (from breast cancer) for a cDNA microarray in which two fluorescent dyes, Cy3 and Cy5, were used for the control and experimental signal of interest. Unfortunately these dyes can be detected by the scanner with different efficiencies. In Figure 1 there is a systematic deviation away from the $y=x$ line as the intensity is increased. A variety of strategies have been proposed to correct for this problem. One of the simplest is to rotate the data clockwise by 45°, determine a one dimensional Lowess curve through the data, correct for any systematic deviation and counter-rotate the data. Alternatives are available such as the use of dye-swapping replicate measurements.

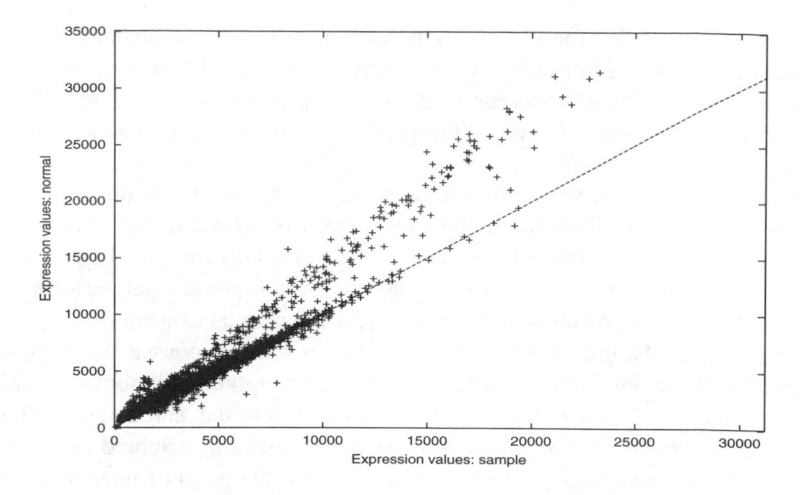

Fig. 1. The different dynamic range of the dyes in this cDNA experiment leads to a systematic deviation with increasing intensity (illustrated by a deviation from the $y = x$ line). This is a relatively mild instance: in some cases the deviation can be substantial.

Having normalised data within-slide, the next step is to normalise the data between slides. The implicit scale and variability within each slide can also make a difference. Again there are a variety of ways to normalise data between slides. For example, housekeeping genes (with expected little variability between slides) could be used to normalise the data. As a simple statistical approach, the median absolute deviation is a robust measure of variability in the data. For a given slide i with median d the median absolute deviation is the median of the differences $|x_{1i} - d|, |x_{2i} - d|, etc$ where x_{1i} is the first measured attribute, x_{2i} is the second, etc. We therefore normalise variability by applying a global rescaling of all the attribute values per slide to standardise the median absolute deviation across all slides.

After image processing there are usually poor readings from some spots on the slide. Suppose the combined dataset is presented with columns corresponding to different samples and rows corresponding to different features. Rows with discardable or missing values could simply be deleted from the subsequent analysis - however, this means we may not use the corresponding gene, and it could be important. If the number of good readings in a row is sufficiently high we could choose to retain the row and impute the small number of poor or missing entries. There is a large body of literature on statistical methods for handling missing values. However, one straightforward method (KNN-impute) is to select a set of *K* features each of which has an expression profile (row) similar to the row containing the missing or discarded entry. Weighting by similarity of expression profile, we calculate and use the weighted average of the K most similar rows as the new entry. Troyanskaya et al [15] report that using the 15 most similar rows gave the lowest error for the range of datasets they considered. Generally it is good practice to flag missing and poor readings and avoid their use in the intra-slide and inter-slide normalisations mentioned above.

3 Classification Techniques

Typically microarray datasets have a large number of features and a small number of examples. Thus, for example, the prostate cancer dataset of Singh et al [13] consists of 50 normal and 52 tumour samples with 12600 features each. Given that the data can therefore be viewed as a sparse set of points in a high-dimensional space (corresponding to a large number of features), it is not surprising that binary class datasets of this type are usually linearly separable: a hyperplane can readily be found which correctly separates both classes. This suggests a preference for simpler algorithms. For example, the perceptron algorithm and its variants can efficiently handle linearly separable problems and can readily be used with these datasets. However, other factors are important. For example, in addition to assigning a class label to a new test point, it would be worth stating a confidence measure too. For some classifiers a confidence or probability measure is given and this will be an advantage in practical applications. A large number of classification algorithms could potentially be used, ranging from discriminant methods, to Gaussian Processes and classification trees. In this section we will only describe three popular choices for illustration: k-nearest neighbours, perceptrons and Support Vector Machines (SVMs).

3.1 K-nearest neighbour classifiers

One of the simplest classifiers to use is k-Nearest Neighbours (kNN). This classifier requires no training and the class of a new point is simply predicted to be the most common class among the k nearest neighbours. In its simplest form and for binary classification with $y_i = \pm 1$, the decision function is:

$$y = sign\left(\sum_{i|x_i \in \mathbf{K}} y_i\right) \tag{1}$$

where $\mathbf{K}$ is the set of neighbours closest to the new point, $\mathbf{x}$. This method can easily be extended to multi-class classification with the class of a new point determined by the consensus of its neighbours.

The set of nearest neighbours is determined by a distance metric that is usually the Euclidean distance in input space. Generally it is best to scale the influence of each neighbour depending on distance from the new point. This is easily accomplished by multiplying the class label of each neighbour by a weighting term, e.g. the reciprocal of the distance between the points. k-nearest neighbours was introduced in 1951 and since then there have been many extensions and variations proposed. One variation is the probabilistic nearest neighbour model (see e.g. [4]): not only does this assign a probability to the predicted label but it also automatically calculates the value of k resulting in a fully autonamous classifier.

3.2 Perceptron classifiers

The perceptron is a simple and effective classifier which can readily handle linearly separable datasets (Fig. 2). Indeed, using the idea of kernel substitution mentioned below, it can also handle non-linearly separable datasets. The *perceptron convergence theorem* states that the perceptron algorithm will converge on a solution after a finite number of passes through the dataset, provided a solution exists. If $\mathbf{x}_i$ are the feature vectors and z are binary-valued outputs then the decision function is:

$$z = sign(\mathbf{w} \cdot \mathbf{x}_i + b). \tag{2}$$

and the learning task is to find a weight vector $\mathbf{w}$ and bias b such that training feature vectors $\mathbf{x}_i$ map correctly to the corresponding labels y_i. This is achieved using an iterative training process with multiple passes through the data using the weight changes:

$$\Delta\mathbf{w} = y_i \mathbf{x}_i H\left[y_i\left(\mathbf{w} \cdot \mathbf{x}_i + b\right)\right] \tag{3}$$

where $H(\theta)$ is the Heaviside Step function, we use $y_i = \pm 1$ and the bias b can be found by adding an extra component (say 0) to the feature vectors fixed at $x_0 = +1$ and with the bias given as $b = w_0$. The weight vector is therefore only corrected by an additive factor $y_i \mathbf{x}_i$ if the decision function gives the wrong label z on presentation of the ith feature vector.

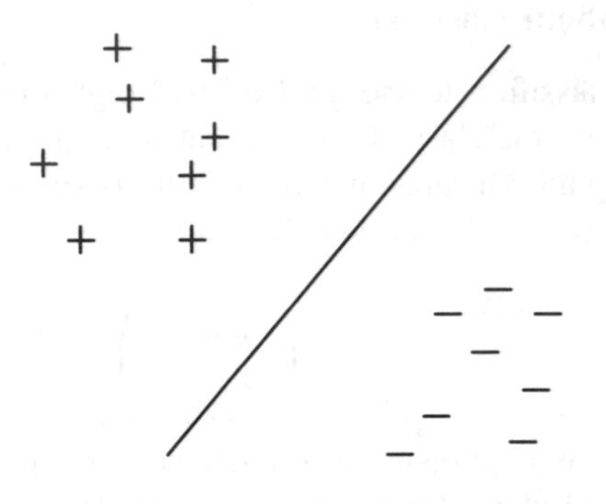

Fig. 2. In input space a linear classifier finds a hyperplane separating the two classes

From (2) we see that the separatrix in the decision function is the linear hyperplane $\mathbf{w} \cdot \mathbf{x} + b = 0$ depicted in Figure 2. In an alternative view (the *geometric dual*) hyperplanes become points and points become hyperplanes (Figure 3). This alternative representation gives insights about the solution found for both the perceptron and SVMs. In this dual representation we have a *version space* inside which each point represents a possible hyperplane correctly separating the datapoints in the original input space (Figure 2). The boundaries of version space derive from the datapoints. The iterations of the perceptron algorithm can be viewed as a trajectory which terminates inside version space starting from outside if the initial weights do not classify the training data correctly. Unfortunately, this indicates that the solution can be biased since it can depend on the starting point and the order of presentation of the patterns in (3). In addition, there is no simple way of implementing a confidence measure for class prediction with the perceptron. Next we will describe SVM classifiers which have a unique solution independent of the order of presentation - in addition a confidence can be assigned to the class label.

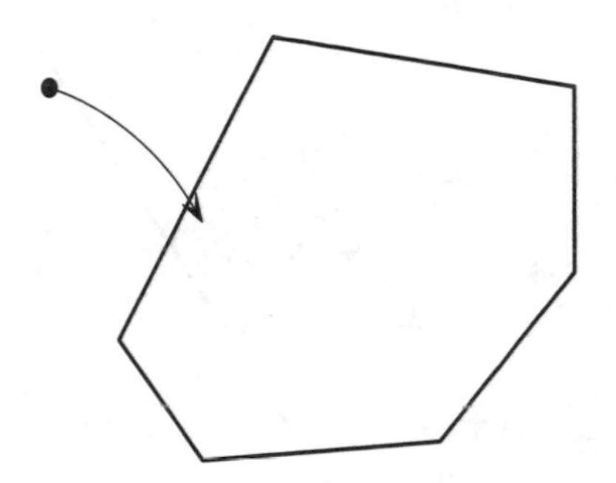

Fig. 3. For the dual representation of Figure 2 datapoints become hyperplanes and hyperplanes correctly separating the data become points inside the *version space* (pictured, version space can be open in general). Thus the iterations of the perceptron algorithm corresponds to a trajectory into version space (shown) where the algorithm terminates with zero training error.

3.3 Support Vector Machines

Support Vector Machines (SVM's) [25, 2, 3] have also been used extensively for classification of microarray data. For binary classification, the motivation for this approach comes from theorems in learning theory which show that good generalisation does not depend on the dimensionality of the space (the number of features used) but on maximising the *margin* or closest distance between the separating hyperplane and closest points on both sides (these are the *support vectors*, Figure 4). For m vectors $\mathbf{x}_i$ each with n features and corresponding labels y_i the task of maximising the margin is provably equivalent to maximising the following function with respect to the parameters α_i

$$W(\alpha) = \sum_{i=1}^{m} \alpha_i - \frac{1}{2} \sum_{i,j=1}^{m} \alpha_i \alpha_j y_i y_j (\mathbf{x}_i \cdot \mathbf{x}_j) \tag{4}$$

subject to the constraints

$$\alpha_i \geq 0 \qquad \sum_{i=1}^{m} \alpha_i y_i = 0. \tag{5}$$

This is a standard constrained quadratic programming problem and can therefore be solved using any one of a number of readily available packages. Since quadratic programming is involved there are no local minima and the learning process always converges to the global minimum (the unique solution described above). For class prediction with microarrays one advantage is that learning is dependent on the number of samples m and not on the number of features (generally the larger number), hence learning is rapid.

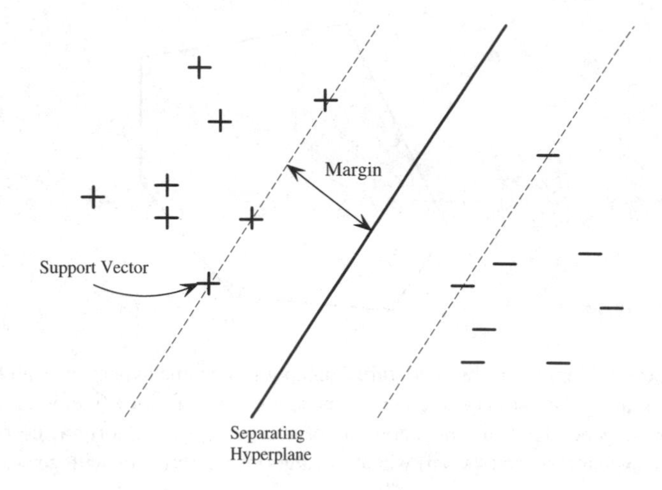

Fig. 4. The SVM solution amounts to finding a separating hyperplane with maximal distance ($margin$) between itself and the closest points of each class on both sides (these are the *support vectors*).

The weight matrix in (2) is given by $\mathbf{w} = \sum_{i=1}^{m} \alpha_i y_i \mathbf{x}_i$, and if the values of α_i are determined at the optimum of (4,5) the decision function is:

$$f(\mathbf{z}) = sign\left(\sum_{i=1}^{m} y_i \alpha_i (\mathbf{x}_i \cdot \mathbf{z}) + b\right) \tag{6}$$

on presentation of a new input $\mathbf{z}$, and where the bias b is determined from:

$$b = -\frac{1}{2}\left[\max_{\{i|y_i=-1\}}\left(\sum_{j=1}^{m} y_j \alpha_j (\mathbf{x}_i \cdot \mathbf{x}_j)\right) + \min_{\{i|y_i=+1\}}\left(\sum_{j=1}^{m} y_j \alpha_j (\mathbf{x}_i \cdot \mathbf{x}_j)\right)\right]. \tag{7}$$

In the solution only some of the α_i values will be non-zero: these correspond to support vectors. Points further away have no influence on the orientation of the separating hyperplane, hence $\alpha_i = 0$ and they make no contribution in the decision function. The spectrum of α-values also carries information about the importance of particular datapoints in the training set, hence enabling *data cleaning*. In particular a large α_i relative to the others can indicate an outlier with undue influence on the orientation of the separating hyperplane. For an early leukaemia microarray dataset

[8] a mislabeled datapoint was detected this way, with subsequent correction of the diagnostic category.

Non-separable datasets. SVMs (and perceptrons) can readily handle non-separable datasets by *kernel substitution*. For SVMs good generalisation does not depend on the dimensionality of the space so mapping non-linearly separable data to a higher dimensional space (called *feature space*) can be implemented without loss of predictive ability. In a higher (or infinite) dimensional space linear separability by a hyperplane can be achieved. In the objective function (4) the datapoints, $\mathbf{x}_i$ only appear inside an inner product. Thus the mapping into feature space involves a mapping of the inner product:

$$\mathbf{x}_i \to \phi(\mathbf{x}_i) \qquad \text{therefore} \qquad \mathbf{x}_i \cdot \mathbf{x}_j \to \phi(\mathbf{x}_i) \cdot \phi(\mathbf{x}_j) \tag{8}$$

and feature space must necessarily be an *inner product* or *Hilbert* space. It is not necessary to define the functional form of the mapping $\phi(\mathbf{x}_i)$ as it is implicitly defined by the choice of mapped inner product or *kernel function*: $K(\mathbf{x}_i, \mathbf{x}_j) = \phi(\mathbf{x}_i) \cdot \phi(x_j)$. Only certain choices for the kernel function are allowed (*Mercer's conditions* must be satisfied [25, 2]). One example of a possible kernel function is the RBF kernel:

$$K(\mathbf{x}_i, \mathbf{x}_j) = \exp\left\{ \frac{-||\mathbf{x}_i - \mathbf{x}_j||^2}{2\sigma^2} \right\} \tag{9}$$

Given the choice of kernel function, learning a dataset for binary classification amounts to finding α_i which maximise the objective function:

$$W(\alpha) = \sum_{i=1}^{m} \alpha_i - \frac{1}{2} \sum_{i,j=1}^{m} \alpha_i \alpha_j y_i y_j K(\mathbf{x}_i . \mathbf{x}_j) \tag{10}$$

subject to the constraints (5). Once the optimal solution has been found, the decision function for a new point $\mathbf{z}$ is given by the sign of

$$f(\mathbf{z}) = \sum_{i=1}^{m} y_i \alpha_i K(\mathbf{z}, \mathbf{x}_i) + b \tag{11}$$

If there is no mapping to feature space then (4,5) apply and we have a *linear kernel*.

Multi-class classification. So far we have discussed binary classification, however, some tasks will involve multi-class prediction. A number of methods have been proposed for performing multi-class prediction with SVMs but few are demonstrably better than using a set of 'one-against-all' classifiers (this approach can be used for perceptrons too). 'One-against-all' classifiers have the drawback that several members of the set may respond to a given input. As an alternative it is possible to use a series of binary classifiers in a tree (Fig. 5) to determine the outcome, and this method works satisfactorily if the number of classes is small.

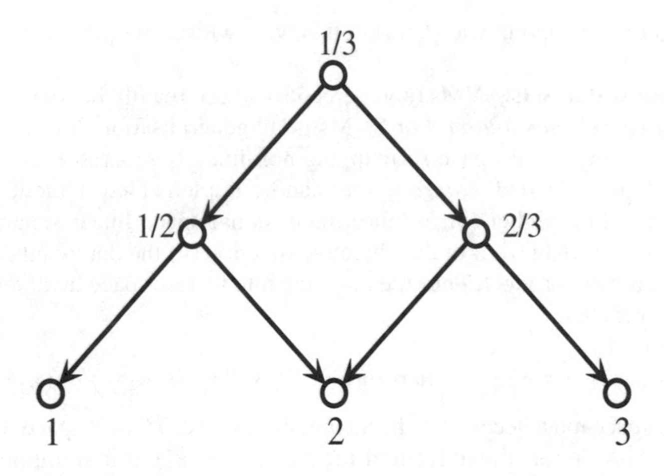

Fig. 5. A multi-class classification problem can be reduced to a series of binary classification tasks (represented by the top two levels in the tree).

Soft Margins. Microarray datasets are typically noisy. An SVM could potentially fit the data well, including the noise, leading to a decrease in predictive accuracy. Noise can be handled using a *soft margin* and two methods are possible. For an L_1-*soft margin*, the optimisation task (4,5) is the same as before except the constraint $\alpha_i \geq 0$ is replaced by $C \geq \alpha_i \geq 0$ where C is the soft margin parameter, while for the L_2-*soft margin* the optimisation task is (4,5) except a small positive quantity λ is added to the kernel diagonal i.e. $K(\mathbf{x}_i, \mathbf{x}_i) \rightarrow K(\mathbf{x}_i, \mathbf{x}_i) + \lambda$. For most microarray datasets the use of a soft margin can lower the test error. However, *a priori* we do not know how much noise is present in the data nor the best value for C or λ, though a numerical cross-validation study (see below) might help to determine the appropriate setting. Without a soft margin and enforcing maximal separation between hyperplane and closest points (Fig. 4) the solution is known as a *hard margin* classifier.

Confidence measures. If used for predicting diagnostic categories, for example, it is useful to have a confidence measure for the class assignment in addition to determining the class label. An SVM with linear kernel does have an inbuilt measure of confidence that could be exploited to provide a confidence measure for the assigned class, i.e. the distance of a new point from the separating hyperplane (Figure 4). A test point a large distance from the separating hyperplane should be assigned a higher degree of confidence than a point which lies close to the hyperplane.

The task of mapping this distance to probabilities has been solved in various ways. Here, we concentrate on one particular method, proposed by Platt [21]. Recall that the output of an SVM, before thresholding to ± 1, is given by

$$f(\mathbf{z}) = h(\mathbf{z}) + b \tag{12}$$

where:

$$h(\mathbf{z}) = \sum_i y_i \alpha_i K(\mathbf{x}_i, \mathbf{z}) \tag{13}$$

We use a parametric model to fit the posterior probability $P(y = 1|f)$ directly. The choice of parametric function for the posterior is the sigmoid (for more details see [21]):

$$P(y = 1|f) = \frac{1}{1 + \exp(Af + B)} \tag{14}$$

with the parameters A and B found from a training set (f_i, y_i). Define t_i as the target probabilities:

$$t_i = \frac{y_i + 1}{2} \tag{15}$$

i.e. using $y_i \in \{-1, 1\}$ we have $t_i \in \{0, 1\}$. We now minimize the log likelihood of the training data:

$$min \left[-\sum_i t_i \log(p_i) + (1 - t_i) \log(1 - p_i) \right] \tag{16}$$

where p_i is simply (14) evaluated at f_i. This is a straightforward 2-dimensional minimisation that can be solved using any one of a number of optimisation routines. Once the sigmoid has been found using this training set, we can use (14) to calculate the probability that a given test point belongs to each class. One question remains: how do we construct a training set to fit the sigmoid. The obvious choice would be the examples from the training set that were used to train the classifier. However, the training process biases the outputs for the support vectors to be ± 1 which is a very unlikely value for a new test point. However, for a linear SVM, the number of support vectors is usually reasonably low and so the bias should not be too severe. Therefore, it is generally acceptable to use the values from the training set to fit the posterior sigmoid. Figure 6 shows the training values and fitted sigmoid from a microarray dataset for an ovarian cancer dataset [12]. Note that no points are present in a large band between $+1$ and -1, due to the use of a hard margin and the data being linearly separable.

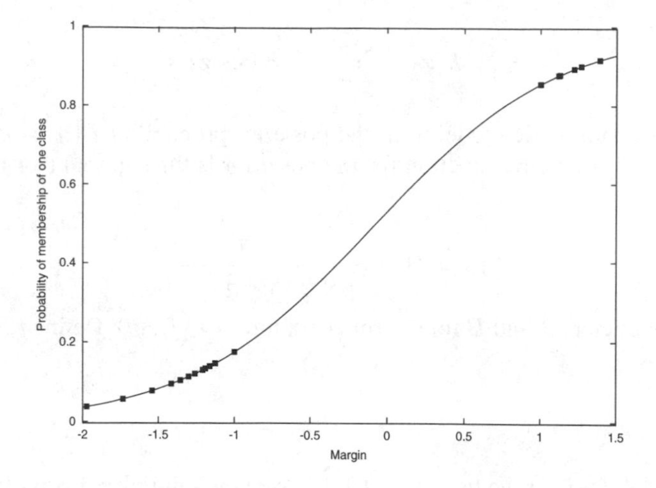

Fig. 6. Probability of membership of one class (y-axis) versus margin. The plot shows the training points and fitted sigmoid for an ovarian cancer data set [12]. A hard margin was used which explains the absense of points in the central band.

3.4 Evaluating the test performance.

With a sparse population of points in a high-dimensional space, for most microarray datasets it is not difficult finding a separating hyperplane and therefore achieving zero training error. Zero training error and separability do not guarantee a low test error (it is possible to construct classifiers which can learn a training set with zero error but exhibit no generalisation ability).

For biological datasets (e.g. for yeast) there may be sufficient data for large training and test sets. However, for medical applications (e.g. cancer) this may not be the case and the paucity of samples can pose a problem. The standard approach to evaluating the test error is to use n-fold cross validation in which the dataset is split into $(n - m)$ training points and m test points with possible multiple resampling to establish a mean and standard deviation for the test error. The limiting case is leave-one-out (LOO) cross-validation with $(n - 1)$ training points and 1 test point, with the single test point rotated successively through the data. LOO cross-validation gives the most *unbiased* estimate of the test error, it is least influenced by individual samples and most representative of performance for the distribution as a whole. However, LOO cross-validation can, depending on classifier, give a low bias but high *variance* solution which will be sub-optimal in terms of generalisation performance (see [22] for a detailed discussion). This *bias-variance tradeoff* can be handled using n-fold cross validation and $n = 5$ and $n = 10$ have been suggested as possible balanced estimates for many classifiers [22].

4 Feature selection

One characteristic of microarray data is that the number of features is usually very large and typically of the order of tens of thousands, often approximating the set of known genes in size. If a broad search is being pursued, the vast majority of these features are likely to be irrelevant for a given classification task and ideally we would like to remove them. Feature selection has two benefits. Firstly, large numbers of irrelevant features effectively inject noise into the classification task and can destroy generalisation, as we will illustrate later with an example. Secondly, from the viewpoint of interpretation, feature selection also highlights the most relevant features or genes in the data. Two general approaches may be used: *filter methods* in which features are scored individually (e.g. using statistical methods) prior to use of the classifier, and *wrapper methods* in which the algorithm uses an internal procedure to eliminate redundant features.

4.1 Filter Methods

The choice of filter method can amount to a prior assumption about the way in which the significance of individual features are ranked. Roughly speaking, filter methods can be viewed as falling into two groupings - those measures more influenced by the consistency of the difference between classes and those more influenced by magnitude differences. For example, the set of ratios (1.1,1.1,1.1,1.1) is consistently different from the set (0.9,0.9,0.9,0.9) and features in both can be separated by a simple threshold (1.0). On the other hand there is a significant difference in the means of the set of ratios (1.0,1.0,5.0,5.0) and (1.0,1.0,0.1,0.1) even though the first two members of each set are the same. As for classification algorithms there are a large number of methods which can be employed. In this section we will therefore describe three commonly used approaches: the Mann-Whitney test and the TNoM score, the Fisher and Golub scores and the t-test (which also comes in many variants). To give an impression of their performance we will also evaluate these scores on a new dataset from cancer research.

Scoring by ranking. Statistical scoring based on ranking will be most influenced by consistency of a difference since if all values belonging to one class are ranked higher than all members of the other class this is determined as an improbable event even if the means of both classes do not differ a great deal. For example, for binary classification, the Mann-Whitney U test provides a measure of the difference between the medians of two populations [5] based on ranking. For a particular feature, two populations are determined from the expression ratios of the two distinctly labelled sample sets. The aim of the test is to estimate the probability that the two populations were drawn from an identical distribution. If this probability is very low, the feature is consistent with the class labels and will be significant. We start by combining the samples from the two classes and ranking them in numerical order. Each sample is then ranked with a value equal to its position in a line (i.e. a rank between 1 and $(n_1 + n_2)$ where n_1 and n_2 are the number of samples in classes 1 and 2.). If expression ratios are tied, they are given the average rank. The ranks for each of the two sample sets are summed. If the sums are R_1 and R_2, we now calculate the U statistic from $U_1 = n_1 n_2 + 0.5 n_1(n_1+1) - R_1$ and $U_2 = n_1 n_2 + 0.5 n_2(n_2+1) - R_2$. Choosing the smaller of U_1 and U_2 as the test statistic U we calculate $z = (U - \mu_U)/\sigma_U$ using:

$$\mu_U = \frac{n_1 n_2}{2} \qquad \sigma_U = \sqrt{\frac{n_1 n_2 (n_1 + n_2 + 1)}{12}} \tag{17}$$

z is a normally distributed random variable ($\mathcal{N}(0,1)$) and hence it is straightforward to determine the probability: the lower the probability the more significant the feature for discriminating the two classes.

Designed specifically for microarray datasets the Threshold Number of Misclassifications (TNoM) score [1] is closely related to ranking scores and will give a similar feature ranking to the Mann Whitney test, for example. We calculate the TNoM score by finding the best classification performance possible for the given feature. For the m values of the feature x_i (components of the feature vector) and label y_i we evaluate:

$$TNoM = \min_{a,b} \sum_{i=1}^{m} H\left[y_i\left(ax_i + b\right)\right] \tag{18}$$

thus we count an error every time $y_i \neq sign(ax_i + b)$. Since the argument inside the Heaviside step function is invariant under an arbitrary positive rescaling we can set $a = 1$ and evaluate the score using a 1-dimensional minimisation on b. If we are able to find a threshold such that all values above it belong to one class and all below belong to the other, then the TNoM score is 0. If the best threshold involves one misclassification the TNoM score is 1, etc. Given a particular class distribution we can also calculate the probability of getting a feature with any particular TNoM score. We can then compare the cumulative distribution function of the TNoM scores for a microarray dataset with the theoretical scores for a null model (i.e. purely random data) and thus determine if there is a significantly high number of low (i.e. good) scoring features compared to random occurrence and therefore if the data carries a significant information load. For example, in Figure 7 we give a plot of the cumulative distributions for an ovarian cancer data set [12]. Here, the curve determined from the ovarian cancer dataset lies considerably to the left of the theoretical curve expected for purely random data, suggesting a large number of discriminating features and significant information content.

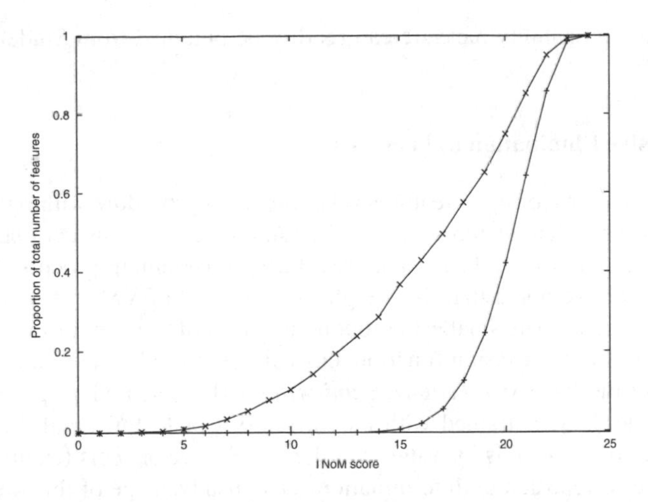

Fig. 7. TNoM cumulative score curves for an ovarian cancer dataset [12]. The x-axis is TNoM score and y axis is the proportion of the total number of features. The left-hand curve corresponds to actual scores and the right-hand curve to the theoretical score for a null (random) model. As the curve from the actual scores lies significantly to the left this means the data contains a significant information content.

The Fisher and Golub scores. The second family of feature scoring techniques has more of an emphasis on the differences in magnitudes of the expression values between classes. Thus if we derive the means and standard deviations for the samples in each class, a good discriminating feature would have a large separation between the means and small standard deviations. If we define the means of the samples in the two classes as μ_1 and μ_2 and their standard deviations as σ_1 and σ_2, the Fisher (F) and Golub (G) scores are defined as follows

$$F = \frac{(\mu_1 - \mu_2)^2}{\sigma_1^2 + \sigma_2^2} \qquad G = \frac{|\mu_1 - \mu_2|}{\sigma_1 + \sigma_2} \tag{19}$$

Scoring using the t-test. The final score we will consider is the well-known t-test for the difference between means of two populations. We calculate the means and variances (μ_1, μ_2, σ_1^2 and σ_2^2) for the two classes and then a weighted average of the two variances, with

$$s^2 = \frac{(n_1 - 1)\sigma_1^2 + (n_2 - 1)\sigma_2^2}{n_1 + n_2 - 2}. \tag{20}$$

and a test statistic t using

$$t = \frac{\mu_1 - \mu_2}{s\sqrt{\frac{1}{n_1} + \frac{1}{n_2}}} \tag{21}$$

from which a probability measure can readily be obtained from Student's distribution.

4.2 Recursive Elimination of Features

Rather than a prior scoring of features we could use a procedure within the algorithm to eliminate redundant features. As an illustration we will consider one method for SVMs which removes irrelevant features during the training process. When using a linear kernel, we noted that the weight matrix for an SVM can be expressed as $\mathbf{w} = \sum_{i=1}^{m} y_i \alpha_i \mathbf{x}_i$. The smallest component values of the weight matrix will have least influence in the decision function and will therefore be the best candidates for removal. For the *Recursive Feature Elimination* (RFE) method proposed by Guyon et al [20], the SVM is trained with the current set of features and the best candidate feature for removal is identified via the weight vector. This feature is removed and the process repeated until termination. One disadvantage of this method is that the process will be slow if there are a large number of features (typically the case for modern high density microarrays), though features could be removed in batches, of course. The algorithm can be terminated if the test error is logged throughout and passes through a minimum though, for SVMs, theoretical criteria such as LOO bounds (which estimate generalisation performance) can also be used [24]. This approach is general and can be applied to the weights generated using the perceptron algorithm, for example.

Aside from RFE there are a number of other approaches where feature selection is implemented directly within the algorithm. For example, with Bayesian approaches a *Bayesian prior* can be incorporated into the design of a classifier favouring sparse solutions, i.e. there is an explicit preference for a solution with a very limited number of features (this is known as Automatic Relevance Determination (ARD), see [23]). Bayesian ARD algorithms for this purpose have been developed and work well with microarray datasets [27].

4.3 Feature Selection: A Case Study.

To illustrate the above we will apply these techniques to a new dataset for predicting relapse versus non-relapse for a paediatric malignancy (R.D. Williams *et al.* manuscript in preparation). In this study cDNA microarrays were used with an approximately balanced dataset of 27 samples. The normalisations mentioned in section 2 were very important and without proper intra- and inter-slide normalisation we found little predictive ability. Normalisations were implemented using a pre-existing software package [18] based on the Statistics for Microarray Analysis (SMA) R package of Speed et al [19].

A Support Vector Machine with linear kernel was used with filter methods for the feature selection. Recursive elimination of features was not used because of the large number of features (17790). Given the small size of the dataset leave-one-out (LOO) appears the best first strategy for evaluating the test error. During evaluation of the test error the 26 examples in the training set change with every leave-one-out rotation. Consequently, to derive a fair test statistic, the statistical scores were determined for each of the 27 evaluations on the test point without incorporating it into the computation of the score.

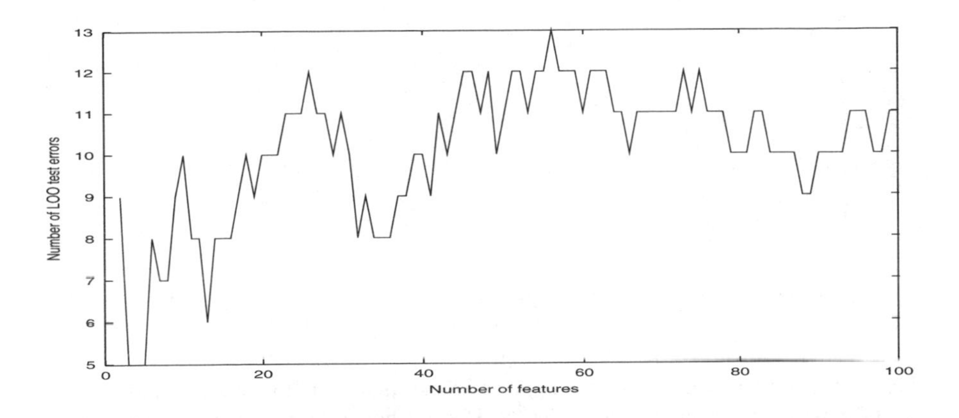

Fig. 8. The number of LOO test errors (y-axis) versus number of top-ranked features (x-axis) remaining with feature-ranking by the Fisher score for predicting relapse or non-relapse. For 4 or fewer features a non-zero training error appears and the test error rises from the minimum of 5/27.

In Figures 8, 9 and 10 we show the LOO test error (y-axis) versus number of top-ranked features (x-axis) remaining for Fisher, Mann-Whitney and t-test scoring of features. All three scores indicate prediction is poor if all the features are used. However, good prediction is achieveable with a small number of features. Thus for the Fisher score the minimal LOO test error is 5/27, for Mann-Whitney 4/27 and for the t-test 1/27. Generally the t-test performs better than non-parametric methods such as the Mann-Whitney test so this comparative performance is to be expected. For 9-fold cross-validation with 1000 random reshufflings of the order we get a $6.4 \pm 1.2\%$ precentage test error with 3 features remaining. These results indicate prediction of relapse can be achieved, though any final confirmation would await evaluation on *de novo* data and biological confirmation of the genes used. If, as here, prediction can be achieved with the expression profile of a small set of genes then the result is interesting but should be viewed with caution. If a set of genes have highly correlated expression then only one member may be needed for building a successful predictor (the rest are effectively redundant), though this gene may not be as significant as some of the others. If, for example, the expression pattern of a particular gene is associated with positional effects (e.g. when a biologically significant overexpressed gene is located in an open chromatin domain, or in a region of the genome that has become duplicated in tumour cells), other genes that are not relevant to tumour outcome may be co-regulated by the same mechanism. Their apparent relevance only derives from their position in this genomic region. Consequently the significance of genes should be evaluated by other methods independently of the feature selection used by the classifier.

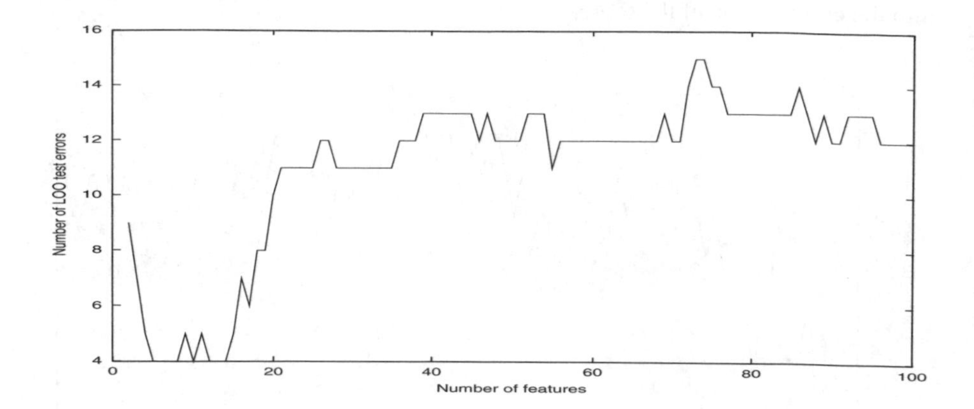

Fig. 9. The number of LOO test errors (y-axis) versus number of top-ranked features remaining (x-axis) with ranking of features by the Mann-Whitney score. The minimum LOO test error improves on the Fisher score in Fig. 8 with a minimum of 4/27.

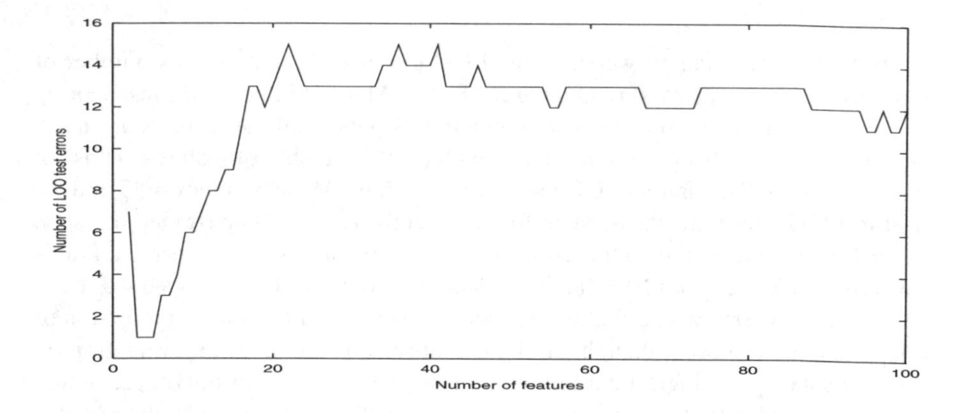

Fig. 10. The number of LOO test errors (y-axis) versus number of top-ranked features remaining (x-axis) with ranking of features by the t-test. The minimum LOO test error was 1 from 27 with 3-5 features remaining. With 2 features remaining a non-zero training error was recorded. The minimal error is lower than for Fisher (Fig. 8) and Mann Whitney (Fig. 9) and the curve is noticeably smoother.

5 Estimating sample size requirements

Suppose for a dataset of size m we evaluate a test error $e(m)$. This test performance may not be adequate for the given task. For example, if the objective is to use a classifier for predicting invasive or non-invasive tumour types the predictor may need to have greater than 95% test accuracy to be acceptable in clinical practice. Hence given the current dataset size and test error we may want to know the dataset size m' which would give a required test error $e(m')$. The answer to this problem lies within learning theory where the theoretical dependence of generalisation error on sample size has been well studied from a number of viewpoints for the perceptron, SVM and other classifiers. *Learning curves* depict this dependence as the sample size varies. The shape of the learning curve depends on the data and the efficiency of the algorithm and in general, for typical datasets, these learning curves have an inverse power-law dependence:

$$e(m) = am^{-\alpha} + b \tag{22}$$

for expected error $e(m)$ given m training samples, *learning rate* a, *decay rate* α and *Bayes error* b, which indicates the minimum test error achievable. The theoretical maximal decay rate is $\alpha = 1$, falling from this optimum for less efficient algorithms or complex learning problems. This functional form is expected to apply to the 25% and 75% quartiles in addition to the mean trend curve. Mukherjee et al [10] have investigated learning curves for a variety of microarray datasets, predominantly for cancer. In Figure 11 and 12 we show numerical test errors (errors on holdout data with resampling) and the learning curves for binary classification using a lymphoma [6] and a colon cancer [11] microarray dataset.

For very small training sets, the above inverse power-law model usually breaks down and there is not enough data to make an accurate extrapolation of the test error. Thus, for a workable method, we must determine the minimum training set size that will give sound results. This is done by finding at what training set size the test error becomes significant when compared with the null hypothesis of a random classifier:

$$H_0 : p(y = 1|x, \{x_1, y_1, \ldots, x_m, y_m\}) = p(y = -1|x, \{x_1, y_1, \ldots, x_m, y_m\}) \tag{23}$$

i.e. the conditional probability of a label being 1 or -1 is equal. The random classifier is constructed by training on the same input data but with the class labels randomly permuted. Suppose we have a total of m data points that we sub-sample into T_1 different training sets of size p and test sets of size $(m-p)$. Now, for each of our T_1 train/test realisations, we construct T_2 random data sets by permuting the training labels. We now use these $T_1 \times T_2$ random datasets to train classifiers and record the error of each classifier when evaluated with the respective non-random test set. Using these errors we construct an empirical distribution function for the random classifier,

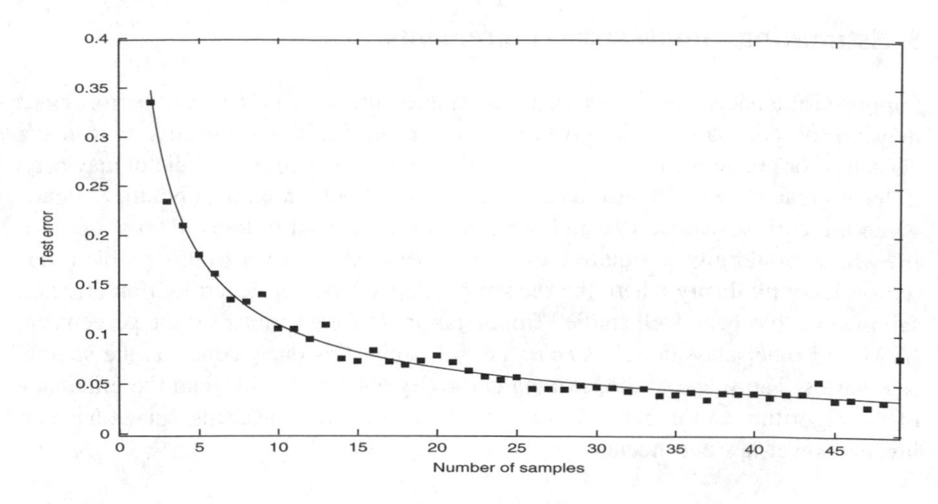

Fig. 11. The learning curve for a lymphoma classification problem [6] distinguishing diffuse large B-cell lymphoma from other types. The fractional error rate is given on the y-axis and the sample size on the x-axis. Both leave-one-out test errors and power law fit are given.

$$P_p^{ran}(x) = \frac{1}{(T_1 \times T_2)} \sum_{i=1}^{T_1} \sum_{j=1}^{T_2} \theta(x - e_{n,i,j}) \tag{24}$$

where $e_{n,i,j}$ is the error of the jth random dataset created from the ith subsampling, with training size p. The significance of the classifier is $P_p^{ran}(\bar{e}_p)$ which is the percentage of random classifiers with error rate smaller than $\bar{e}_p$, the mean error of classifiers trained on the T_1 subsamplings with the *true* labels. So, for example, if it is decided that in order for sample sizes to be used, they should be significant with a probability of 95%, one would simply find the lowest value of p for which $P_p^{ran} > 0.05$. All values of p above this value are deemed to be significant and can be used in fitting the inverse power-law model.

Once a set of p pairs of training set sizes and empirical errors has been gathered, the learning curve can be fitted by minimising:

$$\min_{\alpha,a,b} \sum_{l=1}^{p} (a m_l^{-\alpha} + b - \bar{e}_{m_l})^2 \qquad \alpha, a, b \geq 0. \tag{25}$$

This is a convex optimisation problem when b is fixed. If we fix b, one can estimate α and a by taking logarithms and solving the equivalent linearised minimisation problem:

$$\min_{\alpha,a,b} \sum_{l=1}^{p} (\ln(a) - \alpha m_l + \ln(b - \bar{e}_{m_l}))^2 \qquad \alpha, a, b \geq 0. \tag{26}$$

A straightforward line search procedure over b enables us to find an estimate for α, a and b.

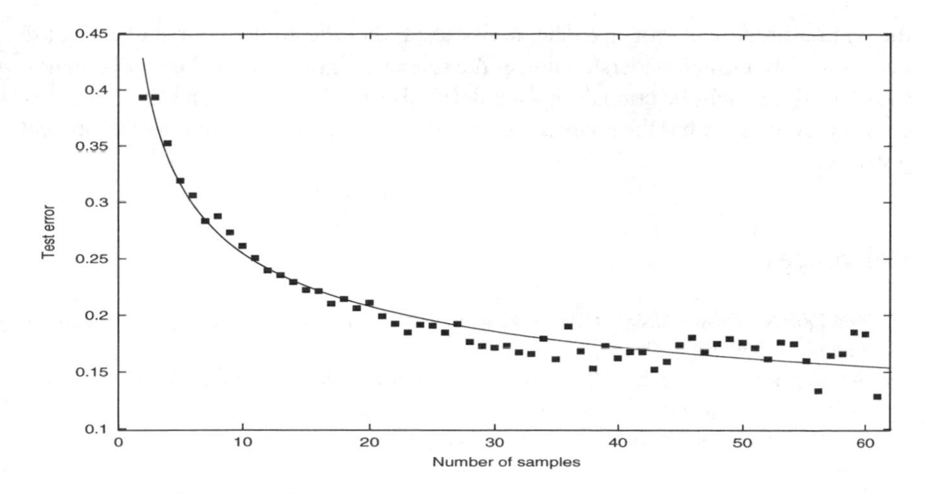

Fig. 12. The learning curve and LOO test errors for a colon cancer classification problem [11].

6 Conclusion

The interpretation of microarray datasets poses new problems for data analysis. It will be important to compare the performance of different classifiers on these datasets, to find the best feature selection strategies, the best method for implementing confidence measures and suitable procedures for handling imbalanced datasets. In addition, the performance of a classifier could be improved by using prior knowledge or the integration of knowledge from other domains. Many other problems arise in this context. For example, a novelty detector might be used to identify new instances which do not apparently belong to any class currently in the data. For some cancer microarray datasets there are samples with little apparent connection to the others, which could have arisen from other locations, belong to functionally different cell types (e.g. stem cells) or be sufficiently abnormal as to be unrepresentative of any assigned classes. In short, a better classifier could be built by not using all the data but by only using prototypical instances with removal of some abnormal datapoints (therefore given as 'unclassifiable'). The best strategies for quality pre-filtering prior to construction of a classifier is an interesting topic for future research.

Aside from classification, microarray datasets prompt many other problems for data analysis. One obvious issue is that the label may be continuous-valued. For example, rather than predicting indolent or aggressive for a tumour sample, it would be better to output a continuous value indicating where on a spectrum the new sample lies. This is the problem of regression and most of the normalisations and feature selection strategies described above will be relevant to this problem also.

The arrival of microarray technology is already giving important insights and this trend will only accelerate in the future. We can expect accurate classifiers which will indicate the detailed sub-type of a disease, the expected future progression and the optimal treatment strategy. The ability to predict the future course of a disease also means an implicit understanding of the role and significance of particular genes. This knowledge will be crucial for drug development. Aside from medical uses, microarray technology has the potential to revolutionise our understanding of biological processes.

References

1. Ben Dor A. Tissue classification of gene expression profiles. *Journal of Computational Biology*, 7:559–583, 2000.
2. Scholkopf B. and Smola A. J. *Learning with Kernels: Support Vector Machines, Regularization, Optimization and Beyond.* MIT Press, 2001.
3. Campbell C. Kernel methods: a survey of current techniques. *Neurocomputing*, 48:63–84, 2002.
4. Holmes C.C. A probabilistic nearest neighbour method for statistical pattern recognition. *Jounal Roy. Statist. Soc. B*, 64(2):295–306, 2002.
5. Rees D.G. *Essential Statistics*. Chapman and Hall, 2001.
6. Alizadeh A.A. et al. Different types of diffuse large b-cell lymphoma identified by gene expressing profiling. *Nature*, 403:503–511, 200.
7. Bhattacharjee A. et al. Classification of human lung carcinomas by mrna expression profiling reveals distinct adenocarcinoma sub-classes. *Proc. Natl. Acad. Sci.*, 98:13790–13795, 2001.
8. Golub T.R. et al. Molecular classification of cancer: Class discovery and class prediction by gene expression monitoring. *Science*, 286:531–537, 1999.
9. Lock C. et al. Gene-microarray analysis of multiple sclerosis lesions yields new targets validated in autoimmune encephalitis. *Nature Medicine*, 8:500–507, 2002.
10. Mukherkee S. et al. Estimating dataset size requirements for classifying dna microarray data. *Journal of Computational Biology*, 10:119–142, 2003.
11. Notterman D. et al. Transcriptional gene expression profiles of colorectal adenoma, adenocarcinoma and normal tissue examined by oligonucleotide arrays. *Cancer Research*, 61:3124–3130, 2001.
12. Schummer M. et al. Comparative hybridization of an array of 21500 ovarian cdnas for the discovery of genes overexpressed in ovarian carcinomas. *International Journal on Genes and Genomes*, 238:375–385, 1999.
13. Singh D. et al. Gene expression correlates of clinical prostate cancer behavior. *Cancer Cell*, 1:203–209, 2002.
14. Sorlie T. et al. Gene expression patterns of breast carcinomas distinguish tumor subclasses with clinical implications. *Proc. Natl. Acad. Sci.*, 98:10869–10874, 2001.
15. Troyanska O. et al. Missing value estimation methods for dna microarrays. *Bioinformatics*, 17:520–525, 2001.
16. Vawter M.P. et al. Microarray analysis of gene expression in the prefrontal cortex in schizophrenia: a preliminary study. *Schizophrenia Research*, 58:11–20, 2002.
17. Yang Y.H. et al. Normalization for cdna microarray data: a robust composite method addressing single and multiple slide systematic variation. *Nucleic Acids Res.*, 30(4):e15, 2002.
18. http://www.maths.lth.se/help/R/com.braju.sma/.
19. http://www.stat.berkeley.edu/users/terry/zarray/Software/smacode.html.

20. Guyon I., Weston J., Barnhill S., and Vapnik V. Gene selection for cancer classification using support vector machines. *Machine Learning*, 46:389–422, 2002.
21. Platt J.C. Probabilistic outputs for support vector machines and comparisons to regularized likelihood methods. In Alexander J. Smola, Peter Bartlett, Schölkopf Bernhard, and Dale Schuurmans, editors, *Advances in Large Margin Classifiers*, 1999.
22. Hastie T. Tibshirani R. and Friedman J. *The Elements of Statistical Learning*. Springer, 2001.
23. Neal R.M. *Bayesian Learning for Neural Networks (Lecture Notes in Statistics 118)*. Springer, 1996.
24. Joachims T. Estimating the generalization performance of a svm efficiently. In *Proceedings of the Seventeenth International Conference on Machine Learning. Stanford, CA.*, 2000.
25. Vapnik V.N. *Statistical Learning Theory*. John Wiley and Sons, inc., 1998.
26. Cleveland W. Robust locally weighted regression and smoothing scatterplots. *Journal of the American Statistical Association*, 74:829–836, 1979.
27. Li Y., Campbell C., and Tipping M. Bayesian automatic relevance determination algorithms for classifying gene expression data. *Bioinformatics*, 18:1332–1339, 2002.

Random Voronoi Ensembles for Gene Selection in DNA Microarray Data

Francesco Masulli[1,2] and Stefano Rovetta[1,3]

[1] INFM-Istituto Nazionale per la Fisica della Materia
Via Dodecaneso 33, I-16146 Genova, Italy
[2] DI-Dipartimento di Informatica, Università di Pisa
Via F. Buonarroti 2, 56127 Pisa, Italy
masulli@di.unipi.it
[3] DISI-Dipartimento di Informatica e Scienze dell'Informazione
Università di Genova, Via Dodecaneso 35, 16146 Genova, Italy
rovetta@disi.unige.it

1 Introduction

1.1 DNA microarrays and the gene selection problem

Currently, cancer and other complex pathologies are analyzed mainly by morphological classification. In the past few decades there have been dramatic improvements, adding many sophisticated methods to the range of available diagnostic tools, but the traditional approaches are still in widespread usage. However, some serious limitations are known to affect these methodologies. For instance, different cancer types and clinical courses, with different response to treatments, can manifest themselves with undistinguishable appearances, not only at morphological inspection, but also from the immunophenotyping, biochemical, and cytogenetic profiles.

This situation is being changed by high throughput data acquisition technologies, of wich a prominent example is the DNA microarray. This tool gives an insight on the molecular aspects of biological phenomena, allowing the monitoring and measurement of expression levels of thousands of genes simultaneously.

One important aspect of disease analysis is the identification of unknown causal relationships. A first step for this task is selecting the relevant variables to be evaluated. This task is known in the pattern analysis field as "input" or "variable selection". Specifically, when analyzing gene expression data, this task is termed *gene selection* and amounts to identifying those genes whose expression level (either higher or lower in pathological tissues than in reference tissues) is more directly related to diagnosis.

1.2 Input selection in pattern recognition

Let's have a closer look at the problem statement. We are given a labeled training sample $\mathbf{x} \in X \subset \mathcal{R}^d$ of n observations. Labels define a dichotomy on X, i.e., the task to be learned is two-class classification. The problem is to assign an importance ranking to each individual input variable x_i with respect to the classification task, with the aim of pointing out which input variables contribute most to the classification performance, and ultimately to select a subset of the most significant input variables.

This problem is properly called *input variable selection*, although it is commonly termed also "feature selection" or even "feature extraction" (which is, more correctly, the task of optimal pre-processing and combining the raw inputs into more significant composite variables).

Variable selection has always been a central problem in pattern recognition. The traditional emphasis has always been on technological issues (enhancing performance of automated recognition methods, lowering computational requirements, reducing the cost of data acquisition, e.g. [1]). However, in relatively recent years, the problem of assessing the relevance of variables has found many applications in basic science.

As noted earlier, DNA microarray data technology provides high volumes of data for each single experiment, yielding measurements for hundreds of genes simultaneously. This is a typical situation in which input selection is required. When inspecting for instance the outcome of a gene expression experiment to identify the "signature" corresponding to a given pathology, the procedure involves almost invariably the application of an automated classification method and the subsequent analysis of the results in seek of the most significant input variables. In this case, input variable selection is a tool supporting scientific inquiry.

In the pattern recognition field there is a vast literature on input selection. One recent contribution is the special issue of the Journal Machine learning Research on this topic [2]; it might be interesting to compare the works presented there with those presented in a previous, related special issue of Artificial Intelligence [3] to see how much the typical size of problems has grown in the years separating the two collections.

In the past few years, for the reasons sketched above, many other contributions come from bioinformatics research. Their aim is to replace traditional variable selection criteria. These are usually very simple, since they rely on the assumption that some variables have *individual* discriminating power; that is, each of them can be used alone to provide a (possibly sub-optimal) classification.

But in general cases there is no theoretical result backing up this assumption, although there are specific applications of pattern recognition where this may be assumed to hold true with good confidence (for instance, in the area of textual document categorization this assumption is known to work quite well). There are cases in which variables, useless by themselves, are useful if taken with other (possibly equally useless) variables. And a clear example of this is provided by the exclusive-OR problem.

Input selection is therefore a matter of identifying the *combinations* of input variables which provide classification performance similar or (in some cases) even better than the whole set of inputs.

The method we propose aims at pinpointing the variables which have the largest influence on the classification performance, also providing a relevance ranking. We are not necessarily interested in finding a good (or optimal) set of variables on which to build a better classifier. We address a so-called *wrapper* approach [4] to supervised variable selection. Wrapper techniques are those relying on the performance of a given learning machine (thus "wrapping" around the learning task). The alternative *filter* approach is based on extracting intrinsic knowledge from the data, by evaluation of some measure of influence of inputs over output such as mutual information [5] analysis of discriminant methods such as logistic regression [6] or simple correlation [7][8]. Finally, we focus on dichotomic (two-class) classification problems.

Given this problem setting, we are interested in obtaining an indication on the possible causes to be included in a more refined model. Therefore in a sense the "selection" phase itself is not even strictly necessary, and we focus on the phase of assessing "input saliency rankings".

The method has been designed for use in typical tasks of analysis of gene expression data (a well known instance of which is represented by [8]), and has been preliminarily validated on actual microarray data.

1.3 Overview of this chapter

The chapter is organized as follows. In Section 2 we review the process of assessing the relative importance of input variables by using derivatives of a discriminating function. Section 3 introduces a method for applying local analysis to the general, non-linear case. In Section 4 through 7 the method is explained in more detail: we discuss its implementation through resampling, which is done by replicating random Voronoi tessellations, and we illustrate how the local analysis is performed with linear Support Vector Machines and how local results are integrated by a suitable clustering algorithm. Section 8 presents experimental results. Finally, Sections 9 and 10 contain discussions and conclusions.

Since the clustering method is original, the Appendix shows how it integrates in a family of clustering algorithms derived by the basic c-Means.

2 Derivative-based Ranking of Input Variables and the Linear Case

Let the input variables x_i be standardized, i.e., $E\{x_i\} = 0\ \forall i$ and $E\{x_i^2\} = 1\ \forall i$. These assumptions can be easily satisfied by pre-processing the input space based on the training set, as in the standard practice. This is especially true of microarray data, where all measurements are made on the same scale and accurate normalization is viewed as a standard part of the preparation of data [9]. Inferring normalization parameters from data with sufficient statistical confidence is not so immediate in general cases where variables are not homogeneous in nature and scale.

Let $r = g(\mathbf{x})$ be the discriminant or decision function, defined on the d-dimensional input vector $\mathbf{x} \in \mathcal{R}^d$ and taking values in $\mathcal{R}$, the discrimination criterion being the value of $y = \text{sign}(r)$. We assume that a good classifier $r = g()$ is given. This is an important assumption. However current classification methods (support vector machines [10]) provide optimal solutions with a minimum of parameter tuning, so that, given a data set, a good classifier is readily obtained.

If we want to analyze what input variables have the largest influence over the output function, we should evaluate the derivatives of r with respect to each variable. This should be done in a neighborhood of the locus $\{\mathbf{x}|g(\mathbf{x}) = 0\}$, and of course requires $g()$ to be locally differentiable (which is a reasonable assumption since smoothing is required by the discrete sampling of data).

This is the so-called *derivative-based saliency*. It is a way to assess the sensitivity of the output to variations in individual inputs. This approach has been used in many contexts and has been experimentally shown to be quite efficient [13].

In the analysis, the following quantities are used:

- The (local) discriminant feature at data point $\overline{\mathbf{x}}$
$$\mathbf{w} = \nabla g(\mathbf{x})|_{\mathbf{x}=\overline{\mathbf{x}}} \tag{1}$$
- The saliency vector
$$\mathbf{t} = \frac{\mathbf{w}}{\max_i \{w_i\}}, \tag{2}$$
where w_i are the individual components of vector $\mathbf{w}$
- The saliency rank vector
$$\mathbf{s} : s_i = \text{rank}\,(t_i, \mathbf{t}), \tag{3}$$
where s_i and t_i are the individual components of vectors $\mathbf{s}$ and $\mathbf{t}$ respectively, and rank (t_i) is the rank of component t_i among the set of component values of vector $\mathbf{t}$.

Given the ranking provided by $\nabla g()$, a variable selection procedure can then be based on a criterion similar to one of the following:

- Fix a number m of input variables and select the first m variables in the ordered list
- Fix a percentage of the total weights and select the inputs which account for that percentage
- Fix a maximum allowed increase in classification error and select the minimum number of variables in the ordered list (starting from the top) for which the error threshold is not exceeded.

The appropriate variable selection strategy depends on the availability of ad-hoc metrics for the applicative problem at hand and also on the problem perspective, since input space reduction aims at the minimum loss of information, while model selection aims at explaining in the clearest way the observed experiments. As a consequence, in the former case bounds on the error will be preferred, while in the latter case the constraint will rather be on the number of inputs.

Let us discuss now the linear case. Early work on classification [14][15] had concentrated on linear classifiers mainly due to computational constraints. In the recent past, linear classifiers have received renewed attention because of their relevance in kernel-based classifier theory and the support vector approach. This justifies the interest of the linear case by itself. Moreover, the linear case can be used to approach nonlinear situations as well, as explained in the following.

In the linear case, $g(\mathbf{x}) = \mathbf{w} \cdot \mathbf{x}$ and

$$\nabla r = \left[\frac{\partial r}{\partial x_1}, \dots, \frac{\partial r}{\partial x_d} \right] = \mathbf{w} \tag{4}$$

In this case, the derivative-based saliency measure can be justified in terms of "percentage of variance explained". The covariance of the input $\mathbf{x}$ has been assumed to be the unit matrix $\Sigma_{\mathbf{x}} = I$. The variance of the output r is therefore $\sigma_r^2 = \mathbf{w}^{\mathsf{T}} \Sigma_{\mathbf{x}} \mathbf{w} = ||\mathbf{w}||_2$. It is clear that, under the assumptions made, the input which gives the largest contribution to the variance of r is the one with the largest coefficient in the vector $\mathbf{w}$. (The assumption above, especially that $\Sigma_{\mathbf{x}} = I$, can be relaxed.)

The single feature r discriminates between the two classes ($r > 0$ and $r < 0$). This feature is given by a linear combination of inputs, with relative weights $\mathbf{w}$. Thus, by sorting the inputs according to their weights, the "importance" ranking is directly obtained.

The mapping from the input space to the discriminant feature r is an orthogonal projection, therefore the selection of the best input variables by evaluation of output sensitivity yields also the projection with minimal error in terms of Euclidean distance (by Luenberger's projection theorem [16]). This justifies the derivative-based approach also from a vector approximation perspective.

3 The Random Voronoi Resampling Method for the General Non-linear Case

In the general non-linear case, it is not possible to define a single clear ranking which holds in any region of the input space. A global approach can employ statistical evaluation of saliency based on data [13], but this requires large datasets which are not generally affordable, and especially so in the case of the DNA microarray methodology.

Our approach involves partitioning the decision function $g()$, and performing local saliency estimates in sub-regions where $g()$ can be approximated with a linear decision function. This local linearization is likely to introduce small errors, due to the local sparsity of data introduced by subsampling.

We can identify three kinds of region: *empty regions* contain no data points; *homogeneous regions* contain points from one class only; *mixed regions* contain points from both classes. These types are all represented in Figure 1.

In the simplest approach, local linearization is made on the basis of an arbitrarily selected partitioning of the data space. Homogeneous and empty regions are discarded. General regions, containing points from both classes, may be crossed by the true decision surface, and in any case a classifier can be built within them; thus they are retained for saliency analysis.

This basic method has several drawbacks:

- Subsampling reduces the cardinality of data (sub)sets, lowering the confidence of classifiers induced on each localized region;
- If the correct decision surface lies between two different localized regions, each of which is homogeneous and has a different class, both regions are discarded and the analysis is distorted by this artifact;
- The number of regions is to be selected a priori, but there is no clear way to decide it;
- The saliency rankings obtained in one region may or may not be in agreement with those in neighboring or other regions, but in most cases they will agree only in part, and there is no way to decide whether several rankings should be combined or kept distinct.

The Random Voronoi Resampling Method we describe in this chapter addresses all these issues. Its main steps, depicted in Figure 2, is detailed in Sections 4 to 7.

4 Random Voronoi Sampling

Tessellations are a geometrical entity useful in disparate areas of science and technology. From telecommunications to three-dimensional surface analysis and reconstruction, from modeling of crystal lattices to analysis of nonlinear dynamics, tessellations are the basic tool for many applications. An example of tessellation of the plane is presented in Figure 1.

A very popular kind of tessellation is represented by Voronoi diagrams [17]. A Voronoi diagram is a tessellation defined by a set of reference points (*sites*); for each site, the corresponding *cell* is the locus of all points in the data space which are closer to that site than to any other site. This property is exploited for instance in multidimensional data compression systems (*vector quantization*).

A Voronoi partition is induced by drawing a Voronoi diagram in the data space. This special kind of partition is defined in an inexpensive way by simply defining the sites. We can obtain a random Voronoi partition by throwing a set of random points in the data space.

This is likely to generate many empty regions, since there is no guarantee that points in the data set will be spanned by random sites. As a consequence, we propose that the random diagram is initialized by a rough vector quantization step, to ensure

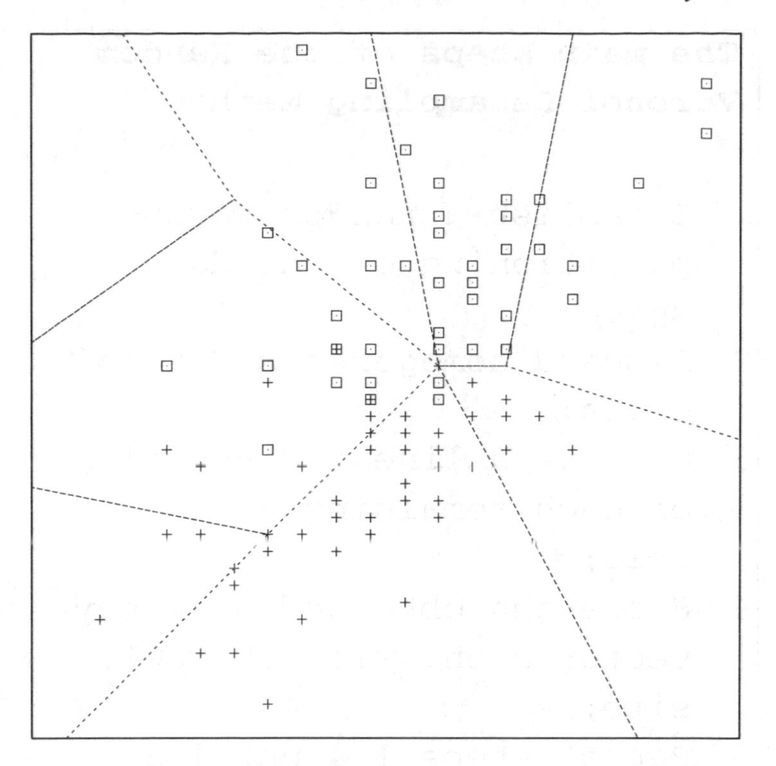

Fig. 1. An arbitrarily partitioned dataset showing empty regions, regions with few samples, homogeneous neighboring regions of different classes.

that sites are placed within the support of the data set. Subsequent random partitions are obtained by perturbation of the initial set of points.

5 Local Linear Classification

Within each Voronoi region, a linear classification is performed. There are many options for analyzing linear separability within a region. The state-of-the-art method is Support Vector Machines (SVM) [10] with a linear kernel.

Given a set of N examples $\mathbf{x}_i$ with their target $y_i = -1, +1$, the Support Vector Machine (SVM) approach consists in mapping the data in a higher dimensional feature space, where we search for a hyperplane separating the two classes. The hyperplane is defined by

$$f(\mathbf{x}) = \mathbf{w}^* \cdot \Phi(\mathbf{x}) + b^* \tag{5}$$

where $\Phi(\mathbf{x})$ is a predefined kernel function (mapping data into the so-called feature space), and the decision rule is given by $sign[f(\mathbf{x})]$. Searching for a decision function $f(\mathbf{x}_i) = y_i$, if this function exists, canonically implies

The main steps of the Random Voronoi Resampling Method

1. Establish a random Voronoi partitioning of the data space;
2. Discard homogeneous Voronoi cells;
3. Compute a linear classifier on each remaining Voronoi cell;
4. Store the obtained saliency vector along with the cell site;
5. Repeat steps 1-4 until a sufficient number of saliency vectors are obtained;
6. Perform joint clustering of the saliency vectors and cell centers;
7. Retrieve cluster centers and use them as estimated local saliency rankings.

Fig. 2. The Random Voronoi Resampling Method

$$y_i(w \cdot \Phi(\mathbf{x}_i) + b) \geq 1, \qquad i = 1, \ldots, N. \tag{6}$$

Often, however, this separating hyperplane does not exist because the data are not linearly separable even in the feature space. To account for non separable data the so-called slack variables

$$\zeta_i \geq 0, \qquad i = 1, ..., N \tag{7}$$

are introduced, to get

$$y_i(w\dot{\Phi}(x_i) + b) \geq 1 - \zeta_i, \qquad i = 1, ..., N. \tag{8}$$

This approach is based on the Structural Risk Minimization (SRM) principle [11]. Rather than minimizing the empirical risk, the estimate of generalization based on training set error (as done in traditional approaches), this principle implies minimizing the probability of misclassification as directly estimated from the theoretical generalization ability of the classifier. This estimate is based on controlling a capacity parameter of the classifier, the so-called *Vapnik-Chervonenkis dimension*. The capacity is in turn directly related to the *classification margin* of the classifier, i.e., the distance from the separating surface to the closest data points.

Therefore, according to the SRM principle, the optimal hyperplane is the one that minimizes a cost function that expresses a combination of two criteria: margin maximization and training error minimization. It is defined as:

$$\Psi(\mathbf{w}, \zeta) = \frac{1}{2}\|\mathbf{w}\|^2 + C\sum_{i=1}^{N}\zeta_i \tag{9}$$

subject to the constraints defined by Eq. (8), where C is a regularization parameter.

The above optimisation problem can be reformulated introducing Lagrange multipliers α_i and solving its dual problem. The final result is expressed as a function of the data in the original data space:

$$f(\mathbf{x}) = \sum_{i=1}^{l} \alpha_i \Phi(\mathbf{x}_i) \cdot \Phi(\mathbf{x}) + b^* \tag{10}$$

where the training examples $\mathbf{x}_i$ corresponding to non zero multipliers α_i are the so-called Support Vectors: all the remaining examples of the training set are irrelevant.

In some cases, the calculation of the dot product $\Phi(\mathbf{x}_i) \cdot \Phi(\mathbf{x})$ can be onerous, but can be reduced by using a suitable kernel function K satisfying the conditions of the Mercer's theorem [12] so that it corresponds to some inner product in the transformed higher dimensional feature space

$$K(\mathbf{x}_i, \mathbf{x}) = \Phi(\mathbf{x}_i) \cdot \Phi(\mathbf{x}) \tag{11}$$

A more detailed description of SVM can be found in [10].

SVMs do not suffer from initialization and parameter sensitivity as other more traditional learning classifiers (e.g. perceptrons), and they provide a single parameter to be tuned for trading off strict separation with robust classification (and generalization).

Working with kernel functions has the computational advantage that, once the *Gram matrix* ($K(\mathbf{x}_i, \mathbf{x}_j)$ for all i and j in the training set) has been computed, dependency on the input space dimensionality is lost and the only important parameter is the number of patterns. Therefore, since our application to feature selection is based on subsampling, the computational complexity of SVM training is small.

6 Building the Ensemble: The Resampling Step

Saliency vectors, as computed in (2), are stored along with their respective sites. This retains the locality information associated with each saliency vector.

The whole set of saliency vectors stored during the iterations of the procedure are analyzed, at the end of the run, by applying a clustering step.

Resampling is one of the techniques used to obtain an *Ensemble method* [18]. Ensemble methods work by combining the outcome of many learning machines or many different instances of a learning machine (as in the present case). The subsequent clustering step acts as the integrator, or arbiter: its role is to integrate the individual outcomes and to output a global response.

In this work, we are interested in partitioning the data space and in obtaining localized "experts". One peculiarity of this approach is that the integrator may output a single response, but it may also output a set of combined responses, each specialized on a given region of the data space. The method can be thus viewed as a sort of "ensemble of ensembles", where the learning machine which is replicated by resampling is in turn a committee of local experts.

Resampling is the key step of the method. It ensures that the data set is smoothly covered and contributes to the stability of the outcomes, by averaging the strong statistical fluctuations. In the proposed approach, the random Voronoi subsampling is replicated by randomly perturbating the initial sites. In our experiments, we applied uniform perturbations with amplitude related to the pairwise distances between data points (e.g. by setting the amplitude equal to the maximum distance).

Unfortunately, it is difficult to obtain theoretical guidelines on how many replications are required as a function of the dimension of the data space and on how to compute the perturbations. This is because theoretical results on stability of Voronoi neighbors are available only for low dimensions [19], and typically rely on assumptions related to the dimension (so that they cannot be generalized to other dimensions).

7 Integration of the ensemble results using the Graded Possibilistic Clustering technique

7.1 The Concept of Graded Possibility

Clustering problems are usually stated as the task of partitioning a set of data vectors or patterns $X = \{x_k\}$, $k \in \{1, \ldots, n\}$, $x_k \in \mathcal{R}^n$ by attributing each data point x_k to a subset $\omega_j \subset X$, $j \in \{1, \ldots, c\}$, defined by its *centroid* $y_j \in \mathcal{R}^n$. This attribution is made based on a given distance $d(\cdot, \cdot)$.

In the Appendix we depict some common aspects of some clustering algorithms derived from the basic c-Means: ("hard") c-Means (HCM) [27], entropy-constrained fuzzy clustering by Deterministic Annealing (DA) [22], Possibilistic c-Means with an entropic cost term (PCM-II) [34], Fuzzy c-Means (FCM) [36, 26]. Among them, a very widely used clustering method is the FCM algorithm, a "fuzzy relative" to the simple c-Means technique [27]. FCM defines the ω_j as fuzzy partitions of the data set X. Variations over this basic scheme try to overcome some of its well-known limitations. The *Deterministic Annealing* (or *Maximum Entropy*) approach [22] does not minimize a simple cost term, but a compound cost function which is the sum of a distortion term $\hat{E}$ and an entropic term $-H$. Optimization is done by fixing a constant value for one of the two terms and minimizing the other; then this step is iterated for decreasing values of the constant, until a global optimum is reached. With this technique it is possible to alleviate the false minima problem.

In decision-making and classification applications, algorithms should feature several desirable properties in addition to the basic decision function. For instance, it is often required that in certain configurations a decision is not made (*pattern rejection*), typically in the presence of outliers. This problem is very well-known and well studied (e.g. see [28][29][30]), and is tackled in a convenient way within the framework of soft-computing, fuzzy, and neural approaches [31][32][33].

However, the clustering problem as stated above implies that the outlier rejection property cannot be achieved. This is because the membership values u_{jk} are constrained to sum to 1 (the *probabilistic* model). By giving up the requirement for strict partitioning, and by resorting to a "mode seeking" algorithm, Krishnapuram and Keller proposed the so-called *possibilistic approach* [21][34], where this constraint is relaxed essentially to

$$u_{jk} \in [0, 1] \quad \forall k, \forall j \tag{12}$$

With this model outlier rejection can be achieved, but at the expense of a clear cluster attribution and other computational drawbacks. The same issue of analyzing the membership interactions on a local basis, as opposed to the global effects induced by the probabilistic model, is considered in [35].

It is worth noting that, on one side, the classic membership model (either hard or fuzzy) implements the concept of partitioning a set into disjoint subsets. This is done through the so-called "probabilistic constraint" by setting $\psi(u_{1k}, \ldots, u_{ck}) = \sum_{j=1}^{c} u_{jk} - 1$. Each membership is therefore formally equivalent to the probability that an experimental outcome coincides with one of c mutually exclusive events.

On the other side, the possibilistic approach implies instead that each membership is formally equivalent to the probability that an experimental outcome coincides with one of c mutually *independent* events. This is due to the complete absence of a constraint on the set of membership values ($\psi \equiv 0$).

However, it is possible (and frequent) that sets of events are neither mutually independent nor completely mutually exclusive either. Instead, events can be loosely related. Often this situation can be modeled by a statistical correlation.

Another interesting model for partial information is the concept of *graded possibility*. This perspective is not grounded in statistics, but in the theories of uncertainty and possibility.

The standard possibilistic approach to clustering implies that all membership values are independent. In contrast, the graded possibilistic model assumes that, when one of the c membership values is fixed, the other $c-1$ values are constrained into a given interval contained in $[0, 1]$. Clearly, this situation includes the possibilistic model, and also encompasses the standard ("probabilistic") approach.

7.2 Modeling Graded Possibility

A class of constraints ψ, including the probabilistic and the possibilistic cases, can be expressed as follows:

$$\psi = \sum_{j=1}^{c} u_{jk}^{[\xi]} - 1, \tag{13}$$

where $[\xi]$ is an interval variable representing an arbitrary real number included in the range $[\underline{\xi}, \overline{\xi}]$. This interval equality should be interpreted as follows: there exists a scalar exponent $\xi^* \in [\underline{\xi}, \overline{\xi}]$ such that the equality $\psi = 0$ holds. This constraint enforces both the normality condition and the required probabilistic or possibilistic constraints; in addition, for nontrivial finite intervals $[\xi]$, it implements the required graded possibilistic condition.

The constraint presented above can be implemented in various ways. We suggest the following particular implementation. Its merit is that it accounts for the probabilistic and possibilistic models as limit cases, being therefore a direct generalization of both.

The extrema of the interval are written as a function of a running parameter α, where

$$\underline{\xi} = \alpha \qquad \overline{\xi} = \frac{1}{\alpha} \tag{14}$$

and

$$\alpha \in [0, 1] \tag{15}$$

This formulation includes as the two extreme cases:

- The "probabilistic" assumption:

$$\alpha = 1 \;\; ; \;\; [\xi] = [1, 1] = 1 \;\; ; \;\; \sum_{j=1}^{c} u_{jk} = 1$$

- The "possibilistic" assumption:

$$\alpha = 0 \;\; ; \;\; [\xi] = [0, \infty] \;\; ; \;\; \sum_{j=1}^{c} u_{jk}^{0} \geq 1 \;,\; \sum_{j=1}^{c} u_{jk}^{\infty} \leq 1$$

Each equation containing an interval is equivalent to a set of two inequalities:

$$\sum_{j=1}^{c} u_{jk}^{\alpha} \geq 1 \qquad \sum_{j=1}^{c} u_{jk}^{1/\alpha} \leq 1.$$

This is depicted in Figure 3 ($c = 2$), where the bounds of the feasible regions are plotted for α decreasing in the direction of the arrows.

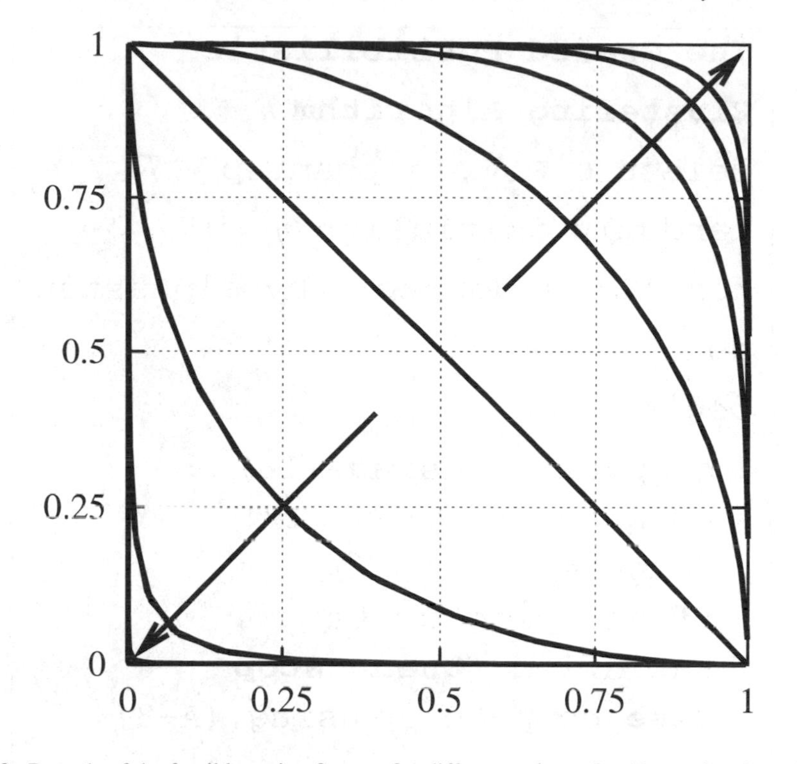

Fig. 3. Bounds of the feasible region for u_{jk} for different values of α (decreasing from 1 to 0 along the direction of the arrows)

In the first limit case, the u_{jk} are constrained on a one-dimensional locus (a line segment). In the second limit case, the locus of the feasible values for u_{jk} is the unit square, which is two-dimensional. In intermediate cases, the loci of feasible values are two-dimensional, but they do not fill the whole square, being limited to eye-shaped areas (increasing with $\alpha \rightarrow 0$) around the line segment.

Another implementation of the interval constraint is used in the outlier rejection application [20]. In this case the upper extremum of $[\xi]$ is fixed to 1 and the lower extremum is α.

7.3 A Graded Possibilistics Clustering Algorithm

In this section we outline a basic example of graded possibilistic clustering algorithm. This is an application of the ideas in the previous section. It is also possible to apply many variations to this algorithm to obtain specific properties.

For the proposed implementation, the free membership function has been selected as in DA and PCM-II:

$$v_{jk} = e^{-d_{jk}/\beta_j}. \tag{16}$$

The generalized partition function can be defined as follows:

```
The Graded Possibilistic
Clustering Algorithm
select c ∈ 𝒩, alphastep ∈ ℛ
randomly initialize y_j
for α = 1 downto 0 by alphastep
do
begin
  compute v_jk using (16)
  compute Z_k using (17)
  compute u_jk = v_jk/Z_k
  if stopping criterion
  satisfied then stop
  else compute y_j using (A-3)
end
```

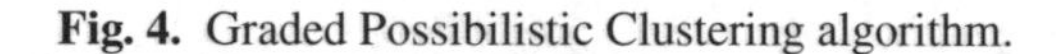

Fig. 4. Graded Possibilistic Clustering algorithm.

$$Z_k = \sum_{j=1}^{c} v_{jk}^{\kappa} \tag{17}$$

where:

$$\begin{array}{lcl} \kappa = 1/\alpha & \text{if} & \sum_{j=1}^{c} v_{jk}^{1/\alpha} > 1 \\ \kappa = \alpha & \text{if} & \sum_{j=1}^{c} v_{jk}^{\alpha} < 1 \\ \kappa = 1 & & \text{else.} \end{array}$$

These definitions ensure that, for $\alpha = 1$, the algorithm reduces to standard DA, whereas in the limit case for $\alpha = 0$, the algorithm is equivalent to PCM-II.

In both cases, β_j can be experimentally estimated (as in PCM) or iteratively "annealed" (as in DA).

Table 1. Relevant inputs for the synthetic problem

Voronoi sites	Saliency vectors	Saliency rank vectors
1	0.91 1.00 0.89 0.79	2 1 3 4
2	0.58 1.00 0.46 0.41	2 1 3 4
	1.00 0.67 0.36 0.51	1 2 4 3
4	1.00 0.41 0.33 0.34	1 2 4 3
	0.30 1.00 0.27 0.27	2 1 3 4
	0.84 0.60 1.00 0.51	2 3 1 4
8	1.00 0.21 0.12 0.19	1 2 4 3
	0.64 1.00 0.25 0.19	2 1 3 4
16	1.00 0.57 0.31 0.33	1 2 4 3
	0.51 1.00 0.44 0.11	2 1 3 4
	0.91 0.88 0.13 1.00	2 3 4 1

7.4 Clustering Saliency Vectors

We notice that the *graded possibilistic model* introduces notable flexibility in the clustering process, while at the same time partial constraints allow the implementation of several desirable properties not attainable with classical approaches, among which there is a user-selectable degree of outlier insensitivity.

In oder to perform the joint clustering of salience vectors and cell centers, we used the Graded Possibilistic Clustering technique [20] to ensure an appropriate level of outlier insensitivity.

The number of cluster centers is assessed by applying a Deterministic Annealing schedule [22] to the resolution parameter β, which is used in the algorithm implementation presented in [20]. The number of clusters is selected to be an arbitrary and abundant quantity at the start of the procedure, when β equals a suitably chosen initial value $\beta^{(i)}$. Cluster centers collapse in early iterations, but with decreasing β they start to differentiate where required by the data distribution. The annealing can stop when β reaches a predefined final value $\beta^{(f)}$, chosen according to a reasonable criterion. For instance, $\beta^{(f)}$ may be comparable to the average pairwise distance between data points.

Table 2. Relevant inputs for the Leukemia data

Gene description	**Gene accession number**	**Correlated class**	**Sign of saliency**
GPX1 Glutathione peroxidase 1	Y00787	AML	–
PRG1 Proteoglycan 1, secretory granule	X17042	AML	–
CST3 Cystatin C (amyloid angiopathy and cerebral hemorrhage)	M27891	AML	–
Major histocompatibility complex enhancer-binding protein mad3	M69043	AML	–
Interleukin 8 (IL8) gene	M28130	AML	–
Azurocidin gene	M96326	AML	–
MB-1 gene	U05259	ALL	+
ADA Adenosine deaminase	M13792	ALL	+

8 Experimental Results

8.1 Results on a Synthetic Dataset

Let's consider now an artificial dataset. Its aim is to verify the properties of the method on a completely known problem.

The four-dimensional data (200 points) have been generated by a mixture of 3 two-dimensional gaussian clusters, one for the first class and the other two for the second class, at the vertices of a triangle. The separating surface between the points of the two classes was therefore approximately hyperbolic. The gaussian mixture data formed the first two components of the input space; the other were generated at random.

Table 1 reports the results for varying number of Voronoi sites. The true relevant components are 1 and 2. Note that in some cases there are clusters in which the values are all close to 1, and the corresponding ranking has no significance. These may be "lost" clusters from the clustering phase, due to a value of the resolution parameter

β that is too small. However, in the majority of cases, only two clusters emerge, and they indicate correctly the two most significant directions for classification.

8.2 Results on a Gene Expression Dataset

The method has undergone validation by comparing its results on the data published by Golub et al. [8]. These well-studied data refer to the analysis, at the molecular level, of two kinds of leukemia, Acute Myeloid Leukemia (AML) and Acute Lymphoblastic Leukemia (ALL). Golub's team was investigating these two forms of cancer with the aid of DNA microarrays. Their aim was to characterize the molecular profile of the two pathologies, trying to discriminate between the two and to find sub-classes among those already known, which could account for different prognostic outcomes in otherwise similar clinical profiles.

This study was one of the most influential for demonstrating the potential of the microarray technology jointly with automated classification methods.

The data were obtained by DNA microarray experiments using high-density oligonucleotide microarray by Affymetrics, reporting on the expression level of 6817 human genes plus controls. Observations refer to 38 bone marrow samples, used as a training set, and 34 samples from different tissues (the test set). The original experimental goals were termed "class discovery" and "class prediction".

In this experiment, we used only the training data for the class discovery (also known as classification) task to discriminate ALL from AML. Classes are in the proportion of 27 ALL and 11 AML observations. The parameters used are as follows: number of sites $= 4$; β decreasing from $\beta^{(i)} = 0.1$ to $\beta^{(f)} = 0.01$ in 10 steps with exponential decay law; perturbation with uniform noise of maximum amplitude 0.5, independent on each input coordinate; 100 perturbations resulting in 400 random partitions of which 61% with mixed classes (the rest being either empty or homogeneous).

The results obtained are summarized in Table 2, which compares the most important genes with those obtained by the original authors. Genes that were indicated both in [8] and by our technique are listed with the sign of the corresponding saliency value.

To test the system, we selected a threshold of 20 top genes in the list found by the final analysis described in Subsection 7. In this way, only the maximum ranks were taken into account. We choose to restrict the analysis to few genes also because a good cluster validation step is not included in the method yet.

Our technique indicates that, among these top 20 genes, 8 of the 50 genes listed in the original work feature the maximum discriminating power. Of the remaining genes, some had less discriminative power and some were not in the list of 50. The results may indicate that, among the 50 most correlated genes found by Golub et al., not all contribute to the actual discrimination to the same extent. In fact, the large number of variables compared to the small number of observations calls for a careful statistical evaluation of the significance of the results obtained.

The ALL class was encoded with $+1$ and the AML class with -1; it is possible to notice that all genes whose expression was found to be correlated with the

"Lymphoblastic" class (ALL) have positive saliency, while those correlated with the "Myeloid" class (AML) have consistently a negative saliency value. Of course, absolute values are not reported since they are not of interest in the present context.

9 Discussion and Open Topics

There are a number of design options and theoretical topics that can be investigated. Some have been touched in the previous sections; here we add some observations.

9.1 Choice of the Scale

The number of Voronoi sites is an important parameter, since it is related to the scale of the tessellation (size of cells). Large cells will tend to contain segments of the separating surface which are difficult to linearize, while small cells will lead to excessively small data subset cardinality, and therefore to low generalization ability.

The selection of the number of sites can be based on estimates of the problem complexity such as those proposed in [23], which are based on geometrical characterization of the data rather than the more usual statistical or information-theoretical consideration. However these must be combined with estimates of generalization to account for the trade-off outlined above.

9.2 Enhancements to the Clustering Step

To make the analysis more robust with respect to variations in the actual saliency values ($\mathbf{t}$), it is possible to analyze the saliency rank values $\mathbf{s}$ instead. Clustering can therefore be made on the space of vectors $\mathbf{s}$.

A given cluster can be analyzed by computing Kendall's rank concordance index W. [24]. Kendall's coefficient for N_c saliency rank vectors $\mathbf{s}^{(1)}, \ldots, \mathbf{s}^{(N_c)}$ is computed as:

$$W = \frac{12 \sum_{c=1}^{N_c} \left(\sum_{i=1}^{d} s_i^{(c)} \right)^2}{N_c(d^3 - d)} - \frac{3(d+1)}{d-1} \tag{18}$$

and is compared to significance tables for W itself or for the related χ^2 statistics.

Clustering can also be modified to incorporate W in its cost function (W within clusters and $(1 - W)$ between clusters) [25]. In principle clustering of rank vectors can be a computationally expensive task.

A cluster validation criterion, added to the clustering phase, can improve the experimental results.

9.3 Enhancements to the Algorithm

There is room for several kinds of optimizations. The technique is especially well suited to parallel implementation at many levels, since the various steps can be pipelined, the subsamples can be processed in parallel, and the Voronoi resampling and clustering phases themselves can be implemented in parallel. All these steps involve very reduced communication. For instance, parallel resampling can be implemented by completely independent random partitions, and communication of subsamples for parallel analysis can be obtained by passing the index of selected patterns. Therefore a Beowulf-type workstation cluster may be proficiently used with limited adaptation effort.

The technique to generate the random perturbations themselves can be optimized, to reduce the number of empty/homogeneous regions, since the data sets are expected to be extremely sparse in the data space. Perturbations can therefore be limited to a subspace, for instance by constraining them to the directions spanned by the versors of the data patterns (e.g., referring to the leukemia data, this is a basis which spans a 38-dimensional subspace of the 6817-dimensional data space). With the subspace optimization, notable performance improvements have been observed.

10 Conclusion

We have described a flexible method for analyzing the relevance of input variables in high dimensional problems. The method has shown the ability to tackle dichotomic problems even in the presence on non-linear separating surfaces. Its behavior has also been validated by comparing the results obtained on a real microarray data set with those published by the original authors.

We have also proposed several open design options and theoretical developments, which are the subject of current research.

Thanks to its ability to tackle non-linear decision surfaces, this method can be used for the exploratory analysis of data from complex phenomena.

Appendix: Unified View of the c-Means Family of Clustering Algorithms

In this Appendix we will propose an unified view of some clustering algorithms derived from the basic c-Means: ("hard") c-Means (HCM) [27], entropy-constrained fuzzy clustering by Deterministic Annealing (DA) [22], Possibilistic c-Means with an entropic cost term (PCM-II) [34], Fuzzy c-Means (FCM) [36].

Let us consider a set of data vectors or patterns $X = \{x_k\}$, $k \in \{1, \ldots, n\}$, $x_k \in \mathcal{R}^n$. On the basis of a given distance $d(\cdot, \cdot)$, clustering will attribute, each data point x_k to a subset $\omega_j \subset X$, $j \in \{1, \ldots, c\}$, defined by its *centroid* $y_j \in \mathcal{R}^n$.

All of the named clustering techniques are based on minimizing the following cost function[1]:

[1] Eq. A-1 includes also FCM, although in the usual formulation this is not evident; see ref. [20]).

$$\hat{E} = \sum_{j=1}^{c} \sum_{k=1}^{n} u_{jk} d_{jk}. \tag{A-1}$$

Here $u_{jk} \in U$ is the degree of membership of pattern x_k to cluster ω_j and $Y = \{y_1, \ldots, y_c\}$. $\hat{E}$ can be termed approximation error, distortion or quantization error, energy, or risk, depending on the application and the nature of the problem. We will refer collectively to these algorithms as the *c-Means (CM) family*.

Miyamoto and Mukaidono [37] show that these algorithms are obtained by adding to the basic cost $\hat{E}$ in (A-1) either regularization terms or the maximum-entropy term

$$-H = \sum_{j=1}^{c} \sum_{k=1}^{n} u_{jk} \log u_{jk} \tag{A-2}$$

which represents the (negative) entropy of the clustering defined by Y, U. We will introduce a formalism to provide an alternative, unified perspective on these clustering algorithms, focused on the memberships u_{jk} rather than on the cost function. We will show that, apart from the possible addition of an entropic term, these algorithms are characterized by specific *feasible regions* for the membership values.

A CM clustering problem is defined by fixing the pair $\{J, \psi\}$, where:

- J is the cost function
- ψ is the constraint on the set of cluster memberships, such that

$$\psi(u_{1k}, \ldots, u_{ck}) = 0 \quad \forall k \in \{1, n\}$$

All the CM algorithms considered define either $J = \hat{E}$ or $J = \hat{E} - H$. Moreover, all the CM algorithms considered require that $u_{jk} \in [0,1]\ \forall j \in \{1, c\}, \forall k \in \{1, n\}$ (*normality* condition). Let v_{jk} be the solution of a CM problem without constraint ψ (formally this can be implemented with $\psi \equiv 0$). We propose to call v_{jk} the *free membership* of pattern x_k in cluster ω_j.

Therefore for all the CM algorithms considered the cluster centroids Y are computed as:

$$y_j = \frac{\sum_{k=1}^{n} u_{jk} x_k}{\sum_{k=1}^{n} u_{jk}} \tag{A-3}$$

characterizing the c-Means principle and therefore the CM family. Memberships are computed as:

$$u_{jk} = \frac{v_{jk}}{Z_k}, \tag{A-4}$$

where Z_k is the (generalized) partition function.

In Table 3 we summarize the main aspects of algorithms belonging to the c-Means (CM) family.

All algorithms are fuzzy techniques, since they adopt the concept of "partial membership" in a set. HCM itself can be cast without imposing the constraint of binary memberships. The relationships among these algorithms are clear from the table.

Table 3. The CM family of clustering algorithms

	J	ψ	v_{jk}	Z_k	*Notes*
DA	$\hat{E} - H$	$\sum_{j=1}^{c} u_{jk} - 1$	$e^{-d_{jk}/\beta}$	$\sum_{j=1}^{c} v_{jk}$	$\beta \in \mathcal{R}$, $\beta > 0$ is the inverse temperature parameter to be increased during the "annealing" process.
PCM-II	$\hat{E} - H$	0	e^{-d_{jk}/β_j}	1	$\beta_j \in \mathcal{R}$, $\beta_j > 0$ are cluster width parameters to be selected a priori before optimization.
FCM	$\hat{E}$	$\sum_{j=1}^{c} u_{jk}^{1/m} - 1$	$1/d_{jk}$	$\left(\sum_{j=1}^{c} v_{jk}^{1/(m-1)}\right)^{m-1}$	$m \in \mathcal{R}$, $m > 1$ is the fuzzification parameter.
HCM	$\hat{E}$	$\sum_{j=1}^{c} u_{jk} - 1$	*See note*	*See note*	v_{jk} and Z_k can be written as for FCM, but their values should be computed for $m \to 1$.

A method to allow for non-extreme solutions is the maximum entropy criterion, which is implemented in the DA and PCM-II algorithms. They are related by the use of the entropic term $-H$, implying a parameter β_j. This parameter is different for each cluster and fixed in PCM-II, while it is constant for all clusters and varying with the algorithm progress in DA.

In the optimization perspective, the parameters β_j arise from the Lagrange multiplier related to the entropic term. They are related to cluster width. In PCM-II their role is crucial, since membership values are not constrained ($\psi \equiv 0$) and are thus allowed to be simultaneously all zero; a means of biasing the solution toward nontrivial values is necessary.

The entropic term in the cost gives rise to free memberships having the functional form

$$v_{jk} = e^{-d_{jk}/\beta_j}, \qquad \text{(A-5)}$$

which characterizes both DA and PCM-II.

An alternative way to obtain non-extreme solutions is introducing nonlinear constraints. The memberships of FCM are equivalent to our $u_{jk}^{1/m}$, rather than u_{jk}. Apart from this constant transformation, our alternative formulation is equivalent and shows that the FCM problem optimizes the same cost function as HCM, but its feasible region is nonlinear (ψ is nonlinear). This allows non-extreme solutions by acting on the membership model.

Acknowledgments

This work was funded by the Italian National Institute for the Physics of Matter (INFM) and by the Italian Ministry of Education, University and Research under a "Cofin2002" grant.

References

1. Sebastiano B. Serpico and Lorenzo Bruzzone, "A new search algorithm for feature selection in hyperspectral remote sensing images", *IEEE Transactions on Geoscience and Remote Sensing*, vol. 39, no. 7, pp. 1360–1367, July 2001.
2. Isabelle Guyon and André Elisseeff, guest editors, "Special issue on variable and feature selection", *Journal of Machine Learning Reesearch*, vol. 3, March 2003
3. D. Subramanian, R. Greiner, J. Pearl, guest editors, "Special issue on relevance", *Artificial Intelligence*, vol. 97, n. 1-2, December 1997
4. George H. John, Ron Kohavi, and Karl Pfleger, "Irrelevant features and the subset selection problem", in *Machine Learning, Proceedings of the Eleventh International Conference, Rutgers University, New Brunswick, NJ, USA*. July 1994, pp. 121–129, Morgan Kaufmann.
5. Nojun Kwak and Chong-Ho Choi, "Input feature selection for classification problems", *IEEE Transactions on Neural Networks*, vol. 13, no. 1, pp. 143–159, January 2002.
6. W. Li and Y. Yang, "How many genes are needed for a discriminant microarray data analysis?", in *Methods of microarray data analysis*, S.M. Lin and K.F. Johnson, Eds., pp. 137–150. Kluwer Academic Publishers, Boston (USA), 2002.
7. G. Grant, E. Manduchi, and C. Stoeckert, "Using non-parametric methods in the context of multiple testing to identify differentially expressed genes", in *Methods of microarray data analysis*, S.M. Lin and K.F. Johnson, Eds., pp. 37–55. Kluwer Academic Publishers, Boston (USA), 2002.
8. T.R. Golub, D.K. Slonim, P. Tamayo, C. Huard, M. Gaasenbeek, J.P. Mesirov, H. Coller, M.L. Loh, J.R. Downing, M.A. Caligiuri, C.D. Bloomfield, and E.S. Lander, "Molecular classification of cancer: Class discovery and class prediction by gene expression monitoring", *Science*, vol. 286, no. 5439, pp. 531–537, October 1999.
9. M. Bilban, L.K. Buehler, S. Head, G. Desoye, and V. Quaranta, "Normalizing DNA microarray data", *Curr Issues Mol Biol*, vol. 2, no. 4, pp. 57–64, April 2002.
10. N. Cristianini and J. Shawe-Taylor, *An Introduction to Support Vector Machines (and Other Kernel-based Learning Methods)*, Cambridge University Press, 2000.
11. V. Vapnik, The Nature of Statistical Learning Theory, Springer, 1995.
12. J.Mercer.Functions of positive and negative type and their connection with the theory of integral equations. Philos. Trans. Roy. Soc. London, A 209:415–446, 1909.
13. Carlo Moneta, Giancarlo Parodi, Stefano Rovetta, and Rodolfo Zunino, "Automated diagnosis and disease characterization using neural network analysis", in *Proceedings of the 1992 IEEE International Conference on Systems, Man and Cybernetics - Chicago, IL, USA*, October 1992, pp. 123–128.
14. Ronald A. Fisher, "The use of multiple measurements in taxonomic problems", *Annual Eugenics*, vol. 7, part II, pp. 179–188, 1936.
15. Frank Rosenblatt, *Principles of Neurodynamics*, Spartan, New York, 1962.

16. David G. Luenberger, *Optimization by Vector Space Methods*, John Wiley and Sons, New York (USA), 1969.
17. Franz Aurenhammer, "Voronoi diagrams-a survey of a fundamental geometric data structure", *ACM Computing Surveys*, vol. 23, no. 3, pp. 345–405, 1991.
18. Thomas G. Dietterich, "Machine-learning research: Four current directions", *The AI Magazine*, vol. 18, no. 4, pp. 97–136, 1998.
19. Frank Weller, "Stability of Voronoi neighborhood under perturbations of the sites", in *Proceedings of Ninth Canadian Conference on Computational Geometry*, 1997.
20. Francesco Masulli and Stefano Rovetta, "The Graded Possibilistic Clustering Model", Proceedings of the International Joint Conference on Neural Networks, Portland, Oregon,IEEE Neural Network Society, Piscataway, NJ, USA, p.p. 791–796, 2003.
21. Raghu Krishnapuram and James M. Keller, "A possibilistic approach to clustering", *IEEE Transactions on Fuzzy Systems*, vol. 1, no. 2, pp. 98–110, May 1993.
22. Kenneth Rose, Eitan Gurewitz, and Geoffrey Fox, "A deterministic annealing approach to clustering", *Pattern Recognition Letters*, vol. 11, pp. 589–594, 1990.
23. Tin Kam Ho and Mitra Basu, "Complexity measures of supervised classification problems", *IEEE Transactions on Pattern Analysis and Machine Intelligence*, vol. 24, no. 3, pp. 289–300, March 2002.
24. Maurice Kendall and Jean Dickinson Gibbons, *Rank Correlation Methods*, Oxford University Press, Oxford (UK), fifth edition, 1990.
25. R. Baumgartner, R. Somorjai, R. Summers, and W. Richter, "Assessment of cluster homogeneity in fMRI data using Kendall's coefficient of concordance", *Magnetic Resonance Imaging*, vol. 17, no. 10, pp. 1525–1532, 1999.
26. James C. Bezdek, *Pattern recognition with fuzzy objective function algorithms*, Plenum, New York, 1981.
27. G.H. Ball and D.J. Hall, "ISODATA, an iterative method of multivariate analysis and pattern classification", *Behavioral Science*, vol. 12, pp. 153–155, 1967.
28. C.K. Chow, "An optimum character recognition system using decision function", *IRE Transactions on Electronic Computers*, vol. 6, pp. 247–254, 1957.
29. C.K. Chow, "An optimum recognition error and reject tradeoff", *IEEE Transactions on Information Theory*, vol. 16, pp. 41–46, 1970.
30. Richard O. Duda and Peter E. Hart, *Pattern Classification and Scene Analysis*, John Wiley and Sons, New York (USA), 1973.
31. Hisao Ishibuchi and Manabu Nii, "Neural networks for soft decision making", *Fuzzy Sets and Systems*, vol. 115, no. 1, pp. 121–140, October 2000.
32. Gian Paolo Drago and Sandro Ridella, "Possibility and necessity pattern classification using an interval arithmetic perceptron", *Neural Computing and Applications*, vol. 8, no. 1, pp. 40–52, 1999.
33. Sandro Ridella, Stefano Rovetta, and Rodolfo Zunino, "K-winner machines for pattern classification", *IEEE Transactions on Neural Networks*, vol. 12, no. 2, pp. 371–385, March 2001.
34. Raghu Krishnapuram and James M. Keller, "The possibilistic C-Means algorithm: insights and recommendations", *IEEE Transactions on Fuzzy Systems*, vol. 4, no. 3, pp. 385–393, August 1996.
35. Antonio Flores-Sintas, José M. Cadenas, and Fernando Martin, "Local geometrical properties application to fuzzy clustering", *Fuzzy Sets and Systems*, vol. 100, pp. 245–256, 1998.

36. J. C. Dunn, "A fuzzy relative of the ISODATA process and its use in detecting compact well-separated clusters", *Journal of Cybernetics*, vol. 3, pp. 32–57, 1974.
37. Sadaki Miyamoto and Masao Mukaidono, "Fuzzy C-Means as a regularization and maximum entropy approach", in *Proceedings of the Seventh IFSA World Congress, Prague*, 1997, pp. 86–91.

Cancer Classification with Microarray Data Using Support Vector Machines

Feng Chu[1] and Lipo Wang[1,2]

[1] School of Electrical and Electronic Engineering, Nanyang Technological Uni-versity, Block S1, Nanyang Avenue, Singapore 639798
[2] Institute of Information Engineering, Xiangtan University, Xiangtan, Hunan, P.R. China

1 Introduction

Microarrays (Schena et al. 1995) are also called gene chips or DNA chips. On a microarray chip, there are thousands of spots. Each spot contains the clone of a gene from one specific tissue. At the same time, some mRNA samples are labeled with two different kinds of dyes, for example,Cy5 (red) and Cy3 (blue). After that, the mRNA samples will be put on the chip and interact with the genes on the chip. This process is called hybridization. After hybridization has finished, the color of each spot on the chip will change. The image of the chip will be scanned out. This image reflects the characteristics of the tissue at the molecular level. If we make microarrays for different tissues, biological and biomedical researchers are able to compare the difference of those tissues at the molecular level. Figure 1 is a description of the process of making microarrays.

An image scanned from microarrays can be considered as an indirect description of the expression levels of the genes, like the “fingerprint" of the genes. Therefore, microarray data are also called gene expression profiles.

One of the most important applications of microarrays is to classify and predict multi-types of cancers. The traditional methods for diagnosis of different types of cancers are mainly based on the morphological appearances of the cancers. However, sometimes it is extremely difficult to find clear distinctions between some types of cancers according to their appearances. Thus, the newly appeared microarray technology is naturally applied to this muddy problem. In recent years, gene-expression-based cancer classifiers have achieved good results in the classifications of lymphoma (Alizadeh et al. 2000), leukemia (Golub et al. 1999), breast cancer (Ma et al. 2003), liver cancer (Chen et al. 2002), and so on.

Gene-expression-based cancer classification is challenging in view of the following two properties of gene expression data. Firstly, gene expression data usually are very high dimensional. The dimensionality usually ranges from several thousands to over ten thousands. Secondly, gene expression data sets usually contain relatively small numbers of samples, e.g., a few tens. If we treat this pattern recognition problem with supervised machine learning approaches, we need to deal with the shortage of training samples and high dimensional input features.

Recent approaches to solve this problem include artificial neural networks (Khan et al. 2001), an evolutionary algorithm (Deutsch 2003), nearest shrunken centroids (Tibshirani et al 2002), a graphical method (Bura and Pfeiffer 2003), etc.

In this chapter, we will use support vector machines (SVMs) to solve the problem of cancer classification based on gene expression data. This chapter is organized as follows. Three gene expression data sets used in this chapter are described in section 2. Then we discuss an important step for dimension reduction, i.e., gene selection, in section 3. The structure and the algorithm of SVMs will be introduced in section 4. Some discussions and conclusions are presented in the last section.

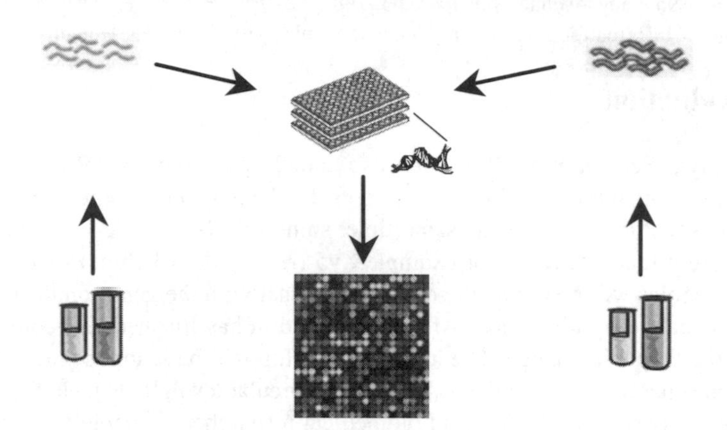

Fig. 1. The process of making microarrays.

2 Gene Expression Data Sets

In the following parts, we describe three data sets to be used in this chapter. One is the small round blue cell tumors (SRBCTs) data set (Khan et al. 2001). Another is the lymphoma data set (Alizadeh et al. 2000). The last one is the leukemia data set (Golub et al. 1999).

2.1 The SRBCT Data Set

The SRBCT data set (Khan et al. 2001) can be obtained from the website http://research.nhgri.nih.gov/microarray/Supplement/. The entire data set includes the expression data of 2308 genes. There are totally 63 training samples and 25 testing samples, five of the testing samples being not SRBCTs. The 63 training samples contain 23 Ewing family of tumors (EWS), 20 rhabdomyosarcoma (RMS), 12 neuroblastoma (NB), and 8 Burkitt lymphomas (BL). And the 20 SRBCTs testing samples contain 6 EWS, 5 RMS, 6 NB, and 3 BL.

2.2 The lymphoma Data Set

The lymphoma dataset (Alizadeh et al. 2000) can be obtained from the website http://llmpp.nih.gov/lymphoma. In this data set, there are 42 samples derived from diffuse large B-cell lymphoma (DLBCL), 9 samples from follicular lymphoma (FL), and 11 samples from chronic lymphocytic lymphoma (CLL). The entire data set includes the expression data of 4026 genes. In this data set, a small part of data is missing. A k-nearest neighbor algorithm was applied to fill those missing values (Troyanskaya 2001).

2.3 The leukemia Data Set

The leukemia data set (Golub et al. 1999) can be obtained at (http://www-genome.wi.mit.edu/cgi-bin/cancer/publications/pub_paper.cgi?mode=view&paper id=43). The samples in this data set belong to two types of leukemia, i.e., the acute myeloid leukemia (AML) and the acute lymphoblastic leukemia (ALL). Among these samples, 38 of them are used for training and the other 34 independent samples are for testing. The entire leukemia data set contains the expression data of 7129 genes.

Ordinarily, raw gene expression data should be normalized to reduce the systemic bias introduced during experiments. For the SRBCT and the lymphoma data sets, the data after normalization can be found on the web. However, for the leukemia data set, normalized data are not available. Thereafter, we need to do normalization by ourselves.

We followed the normalization procedure used in (Dudoit et al. 2002). Three steps were taken, i.e., (a) thresholding: with a floor of 100 and a ceiling of 16000, that is, if a value is greater/smaller than the ceiling/floor, this value is replaced by the ceiling/floor; (b) filtering, leaving out the genes with $\max/\min \leq 5$ or $(\max - \min) \leq 500$, here max and min refer to the maximum and minimum of the expression values of a gene, respectively; (c) carrying out logarithmic transformation with 10 as the base to all the expression values. 3571 genes survived after these three steps. Furthermore, the data were standardized across experiments, i.e., minus the mean and divided by the standard deviation of each experiment.

3 Gene Selection Methods

3.1 Introduction

As we mentioned in the previous part, a typical gene expression data set usually contains expression profiles of a large number of genes. Among all those genes, only a small part may benefit the correct classification of cancers. The rest of the genes have little impact on the classification. Even worse, some genes may act as "noise" and undermine the classification accuracy. Hence, to obtain good classification accuracy, we need to pick out the genes that benefit the classification most. In addition, gene selection is also a procedure of input dimension reduction, which leads to a much less computation load to the classifier. Maybe more importantly, reducing the number of genes used for classification can help researchers put more attention on these important genes and find the relationship between those genes and the development of the cancers.

3.2 Principal Component Analysis

The most widely used technique for input dimension reduction in gene expression analysis is principal component analysis (PCA) (Simon 1999). The basic idea of principal component analysis is transforming the input space into a new space described by the principal components (PCs). All the PCs are orthogonal to each other and are ordered according to the absolute values of their eigenvalues. The k-th PC is the vector with the k-th largest eigenvalue. By leaving out the vectors with small eigenvalues, the dimensionality of the input space is reduced. Because PCA chooses vectors with the largest eigenvalues, it covers the directions with the largest vector variations in the input space. In the directions determined by the vectors with small eigenvalues, the vector variations are very small. In a word, PCA intends to capture the most informative directions. Figure 2a shows the change of eigenvalues in the lymphoma data set. The classification result using PCA as input dimension reduction scheme is given in Fig. 2b. Through comparing Fig. 2b with Fig. 11a, we found that t-test (to be introduced in the latter part) could achieve much better classification accuracy than PCA.

3.3 Class Separability Analysis for Gene Selection

Another frequently used method for gene selection is to measure the class separability (CS) (Dudoit et al.2002). CS is defined as:

$$CS = S_b / S_w \,, \tag{1}$$

where

$$S_b = \sum_{i=1}^{K} \|\mathbf{m_i} - \mathbf{m}\|^2 \,, \tag{2}$$

$$S_w = \sum_{i=1}^{K} \sum_{l=1}^{n_i} \|\mathbf{x_{il}} - \mathbf{m_i}\|^2 \,, \tag{3}$$

$$\mathbf{m_i} = \sum_{l=1}^{n_i} \mathbf{x_{il}} / n_i \,, \tag{4}$$

$$\mathbf{m} = \sum_{i=1}^{K} \sum_{l=1}^{n_i} \mathbf{x_{li}} / \sum_{i=1}^{C} n_i \,. \tag{5}$$

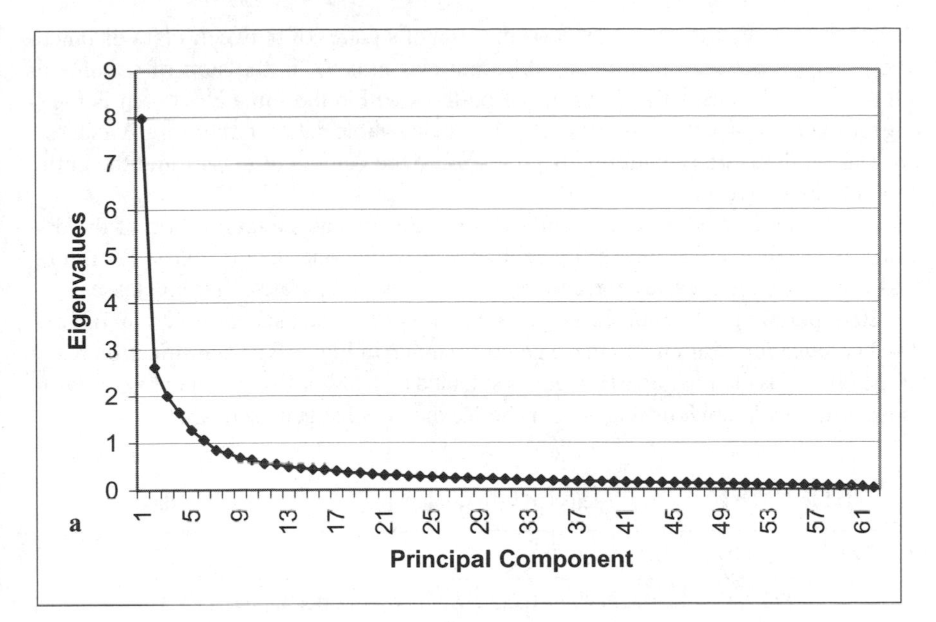

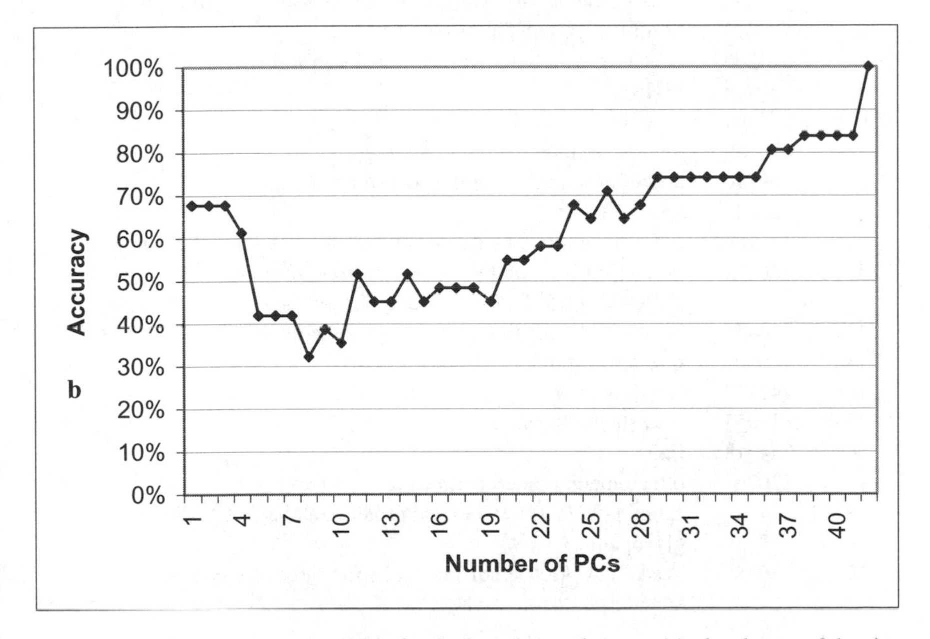

Fig. 2. Principal components analysis for the lymphoma data set. (a): the change of the eigenvalues; (b): the classification result using PCA for input reduction. The horizontal axis is the number of the PCs with the greatest eignvalues that are used.

Here $\|\cdot\|$ is the Euclidean norm. S_b is the sum of squares of between-class distances (the distances between patterns of different classes). S_w is the sum of squares of with-in class distances (the distances of patterns within the same class). $\mathbf{x_{il}}$ is the l-th gene expression pattern in class cancer i. In the whole data set, there are K classes. For class i, there are ni patterns. $\mathbf{m_i}$ and $\mathbf{m}$ are the centers of class i and the entire data set, respectively.

A CS is calculated for each gene. A larger CS indicates a larger ratio of the distances between different classes to the distances within one specific class. Therefore, CSs can be used to measure the capability of genes to separate different classes.

20 important genes with the largest CSs in SRBCT data set are given in Table 1. The classification result using these genes is shown in Fig. 3. The classification result of the same classifier using 20 top genes selected by t-test is also given in Fig. 3. From these results, it is obvious that t-test is better than CS for gene selection.

Table 1. The 20 top genes selected by class separability in the SRBCT data set.

Rank	Gene ID	Gene Description
1	770394	Fc fragment of IgG, receptor, transporter, alpha
2	796258	sarcoglycan, alpha (50kD dystrophin-associated glycoprotein)
3	784224	fibroblast growth factor receptor 4
4	814260	follicular lymphoma variant translocation 1
5	295985	ESTs
6	377461	caveolin 1, caveolae protein, 22kD
7	859359	quinone oxidoreductase homolog
8	769716	neurofibromin 2 (bilateral acoustic neuroma)
9	365826	growth arrest
10	1435862	antigen identified by monoclonal antibodies 12E7, F21 and O13
11	866702	protein tyrosine phosphatase, non-receptor type 13 (APO-1/CD95 (Fas)-associated phosphatase)
12	296448	insulin-like growth factor 2 (somatomedin A)
13	740604	interferon stimulated gene (20kD)
14	241412	E74-like factor 1 (ets domain transcription factor)
15	810057	cold shock domain protein A
16	244618	ESTs
17	52076	olfactomedinrelated ER localized protein
18	21652	catenin (cadherin-associated protein), alpha 1 (102kD)
19	43733	glycogenin 2
20	236282	Wiskott-Aldrich syndrome (ecezema-thrombocytopenia)

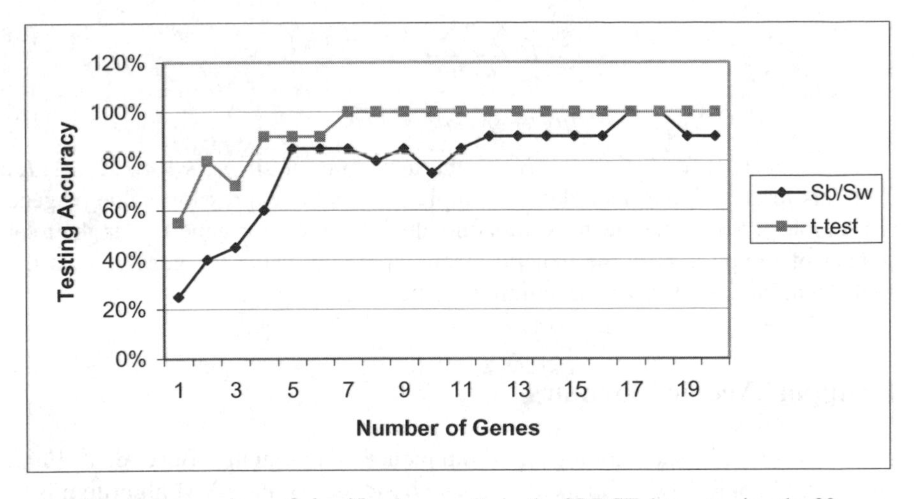

Fig. 3. Comparison of classification results in the SRBCT data set using the 20

3.4 A t-test-based gene selection approach

T-test is a statistical method proposed by Welch (1947). It is to measure how large the difference is between the distributions of two groups of samples. For a specific gene, if it shows larger distinctions between 2 groups, it is more important for the classification of the two groups. To find the genes that contribute most to the classification, t-test has been used in gene selection (Tusher et al. 2001) in recent years.

To select important genes using t-test involves several steps. In the first step, a score based on t-test (named t-score or TS) is calculated for each gene. In the second step, all the genes are rearranged according to their TSs. The gene with the largest TS is put in the first place of the ranking list, followed by the gene with the second greatest TS, and so on. Finally, only some top genes in the list are used for classification.

The standard t-test is only applicable to measure the difference between two groups. Therefore, when the number of classes is more than two, we need to modify the standard t-test. In this case, we use t-test to calculate the degree of difference between one specific class and the centroid of all the classes. Hence, the definition of TS for gene i can be described like this:

$$TS_i = \max\left\{\left|\frac{\bar{x}_{ik} - \bar{x}_i}{m_k s_i}\right|, \ k = 1, 2, \ldots, k\right\}, \tag{6}$$

where

$$\bar{x}_{ik} = \sum_{j \in C_k} \bar{x}_{ij} / n_k, \tag{7}$$

$$\bar{x}_i = \sum_{j=1}^{n} x_{ij} / n, \tag{8}$$

$$s_i^2 = \frac{1}{n-K} \sum_k \sum_{j \in C_k} (x_{ij} - \bar{x}_{ik})^2 \,, \tag{9}$$

$$m_k = \sqrt{1/n_k + 1/n} \,. \tag{10}$$

Here $\max\{y_k,\ k = 1, 2, \ldots, K\}$ is the maximum of all y_k, $k = 1, 2, \ldots, K$. C_k refers to class k that includes n_k samples. x_{ij} is the expression value of gene i in sample j. $\bar{x}_{ik}$ is the mean expression value in class k for gene i. n is the total number of samples. $\bar{x}_i$ is the general mean expression value for gene i. s_i is the pooled within-class standard deviation for gene i.

4 Support Vector Machines

The SVM is a statistical learning algorithm pioneered by Vapnik (Boser et al. 1992; Cortes and Vapnik 1995; Vapnik 1998). The basic idea of the SVM algorithm is to find an optimal hyper-plane that can maximize the margin (a precise definition of margin will be given in the later part) between two groups of samples. The vectors that are nearest to the optimal hyper-plane are called support vectors (vectors with a circle in Fig. 4) and this algorithm is called support vector machine. Compared with other algorithms, SVMs have shown outstanding capabilities in dealing with classification problems. This section briefly describes the SVM.

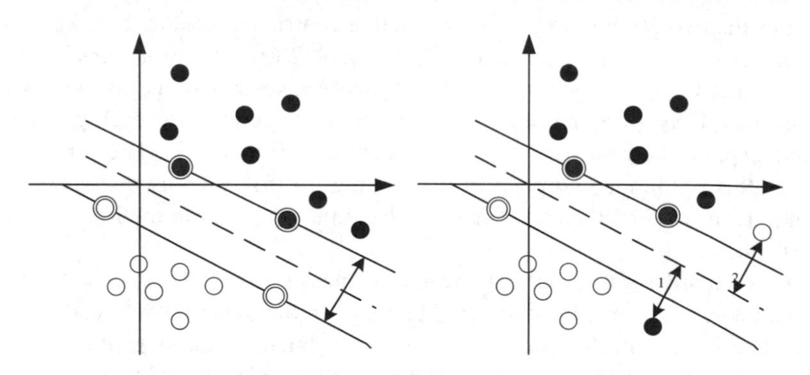

Fig. 4. An optimal hyper-plane for classification in a two dimensional case, for (a) linearly separable patterns and (b) linearly non-separable patterns.

4.1 An Optimal Hyper-plane for Pattern Recognition

4.1.1 Linearly Separable Patterns

Given l input vectors $\mathbf{x}_i \in R^n$, $i = 1, \ldots, l$, that belong to two classes, with desired output $y_i \in \{-1, 1\}$, if there exists a hyper-plane

$$\mathbf{w}^T\mathbf{x} + b = 0 \tag{11}$$

that separates the two classes, that is,

$$\mathbf{w}^T\mathbf{x_i} + b \geq 0, \quad \text{for all } i \text{ with } y_i = +1\,, \tag{12}$$

$$\mathbf{w}^T\mathbf{x_i} + b < 0, \quad \text{for all } i \text{ with } y_i = -1\,, \tag{13}$$

then we say these patterns are linearly separable. Here $\mathbf{w}$ is a weight vector and b is a bias. By rescaling $\mathbf{w}$ and b properly, we can change the two equations above to:

$$\mathbf{w}^T\mathbf{x_i} + b \geq 1, \quad \text{for all } i \text{ with } y_i = +1\,, \tag{14}$$

$$\mathbf{w}^T\mathbf{x_i} + b \leq -1, \quad \text{for all } i \text{ with } y_i = -1\,. \tag{15}$$

Or

$$y_i(\mathbf{w}^T\mathbf{x_i} + b) \geq 1\,. \tag{16}$$

There are two parallel hyper-planes

$$\mathbf{H1} : \mathbf{w}^T\mathbf{x_i} + b = 1\,, \tag{17}$$

$$\mathbf{H2} : \mathbf{w}^T\mathbf{x_i} + b = -1\,. \tag{18}$$

We define the distance ρ between $\mathbf{H1}$ and $\mathbf{H2}$ as the margin between the two classes (Fig. 4a). According to the standard result of the distance between the origin and a hyper-plane, we can figure out that the distances between the origin and $\mathbf{H1}$ and $\mathbf{H2}$ are $|b-1|/||\mathbf{w}||$ and $|b+1|/||\mathbf{w}||$, respectively. The sum of these two distances is ρ, because $\mathbf{H1}$ and $\mathbf{H2}$ are parallel. Therefore,

$$\rho = 2/||\mathbf{w}||\,. \tag{19}$$

Our objective is to maximize the margin between the two classes, i.e., to minimize $||\mathbf{w}||$. This objective is equivalent to minimizing the cost function:

$$\psi(\mathbf{w}) = \frac{1}{2}||\mathbf{w}||^2\,. \tag{20}$$

Then, this optimization problem subject to the constraint (16) can be solved using Largrange multipliers. The Lagrange function is

$$L(\mathbf{w}, b, \alpha) = \frac{1}{2}||\mathbf{w}||^2 - \sum_{i=1}^{l} \alpha_i \left[y_i(\mathbf{w}^T\mathbf{x_i} + b) - 1\right]\,, \tag{21}$$

where $\{\alpha_i,\ i = 1, 2, \ldots, l\}$ are Lagrange multipliers. Differentiating this Lagrange function, we obtain

$$\frac{\partial L(\mathbf{w}, b, \alpha)}{\partial \mathbf{w}} = \mathbf{0}\,, \tag{22}$$

$$\frac{\partial L(\mathbf{w}, b, \alpha)}{\partial b} = 0\,. \tag{23}$$

Considering Wolfe's dual (Fletcher, 1987), we can obtain a dual problem of the primal one: maximize:

$$\text{maximize: } Q(\alpha) = \sum_{i=1}^{l} \alpha_i - \frac{1}{2} \sum_{i=1}^{l} \sum_{j=1}^{l} \alpha_i \alpha_j y_i y_j \mathbf{x_i}^T \mathbf{x_j} \,, \tag{24}$$

subject to:

$$\sum_{i=1}^{l} \alpha_i y_i = 0 \,, \tag{25}$$

$$\alpha_i \geq 0 \,. \tag{26}$$

From this dual problem, the optimal weight vector, i.e., $\mathbf{w}_o$ and the optimal Lagrange multipliers, i.e., $\alpha_{o,i}$ of the optimal hyper-plane can be obtained:

$$\mathbf{w}_o = \sum_{i=1}^{l} \alpha_{o,i} y_i \mathbf{x_i} \,, \tag{27}$$

$$\sum_{i=1}^{l} \alpha_{o,i} y_i = 0 \,. \tag{28}$$

4.1.2 Linearly Non-separable Patterns

If the vectors $\mathbf{x}_i \in R^n$, $i = 1, \ldots, l$, cannot be linearly separated, we would like to slack the constraints described by equation (16). Here we introduce a group of slack variables, i.e., ξ_i.

$$y_i(\mathbf{w}^T \mathbf{x_i} + b) \geq 1 - \xi_i \,, \tag{29}$$

$$\xi_i \geq 0 \,. \tag{30}$$

In fact, ξ_i is the distance between the training example $\mathbf{x_i}$ and the optimal hyper-plane (Fig. 4b). For $0 \geq \xi_i \geq 1$, $\mathbf{x_i}$ falls into the region between the two hyper-planes, i.e., **H1** and **H2**, but in the correct side of the optimal hyper-plane. However, for $\xi_i > 1$, $\mathbf{x_i}$ falls into the wrong side of the optimal hyper-plane.

Since we hope the optimal hyper-plane can maximize the margin between the two classes and minimize the errors, we rewrite the cost function from equation (20) to:

$$\psi(\mathbf{w}, \xi) = \frac{1}{2}||\mathbf{w}||^2 + C \sum_{i=1}^{l} \xi_i \,, \tag{31}$$

where C is a positive factor. We will discuss the functionality of C in the next section. This cost function must satisfy the constraints of (29) and (30). There is also a dual problem:

$$\text{maximize: } Q(\alpha) = \sum_{i=1}^{l} \alpha_i - \frac{1}{2} \sum_{i=1}^{l} \sum_{j=1}^{l} \alpha_i \alpha_j y_i y_j \mathbf{x_i}^T \mathbf{x_j} \,, \tag{32}$$

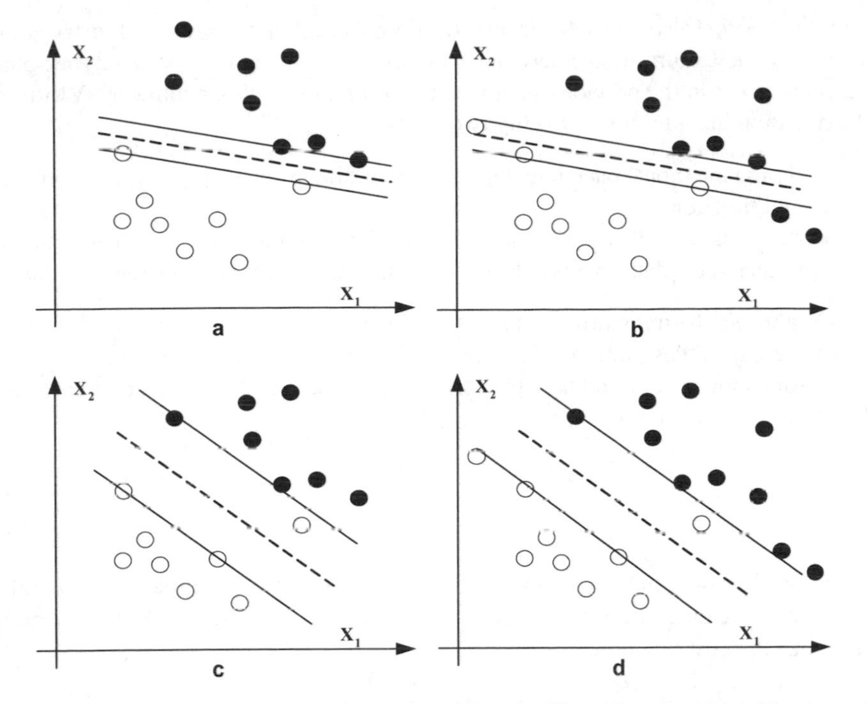

Fig. 5. The influence of C to the performance of the classifier. (a) a classifier with a large C (small margin); (b) an overfitting classifier; (c) a classifier with a small C (large margin); (d) a classifier with a proper C.

subject to:

$$\sum_{i=1}^{l} \alpha_i y_i = 0\,, \tag{33}$$

$$C \geq \alpha_i \geq 0\,. \tag{34}$$

From this dual problem, the optimal weight vector, i.e., $\mathbf{w}_o$ and the optimal Lagrange multipliers, i.e., $\alpha_{o,i}$ of the optimal hyper-plane can be obtained. They are same as their counterparts in (27) and (28), except that the constraints change to (33) and (34).

4.2 A Binary Nonlinear SVM Classifier

According to (Cover 1965), if a nonlinear transformation that can map the input feature space into a new feature space whose dimension is high enough, the classification problem is more likely to be linearly solved in this new high dimensional space. In view of this theorem, the *nonlinear* SVM algorithm performs such a transformation to map the input feature space to a new space with much higher dimension. Actually, other kernel learning algorithms, such as radial basis function (RBF) neural networks, also do such a transformation for the same reason. After the transformation, the features in the new space are classified using the optimal hyper-plane we constructed in the previous sections. Therefore, using this nonlinear SVM to do classification includes the following two steps:

1. Mapping the input space into a much higher dimensional space with a nonlinear kernel function.
2. Performing classification in the new high dimensional space by constructing an optimal hyper-plane that is able to maximize the margin between the two classes.

Combining the transformation and the linear optimal hyper-plane, we formulate the mathematical descriptions of this nonlinear SVM as follows.

We are supposed to find the optimum values of weight vector $\mathbf{w}$ and bias b such that they satisfy the constraint:

$$y_i(\mathbf{w}^T\phi(\mathbf{x_i}) + b) \geq 1 - \xi_i, \quad i = 1, 2, \ldots, l, \tag{35}$$

$$\xi_i \geq 0, \tag{36}$$

where $\phi(x_i)$ is the function mapping i-th pattern vector to a potentially much higher dimensional feature space. The weight vector $\mathbf{w}$ and the slack variables ξ_i should minimize the cost function:

$$\psi(\mathbf{w}, \xi) = \frac{1}{2}\mathbf{w}^T\mathbf{w} + C\sum_{i=1}^{l}\xi_i. \tag{37}$$

This optimization problem is very similar to the problem we have dealt with using a linear optimal hyper-plane. The only difference is that the input vectors $\mathbf{x_i}$ have been replaced by $\phi(\mathbf{x_i})$. In general, this cost function trades off the two goals of the binary SVM, i.e., to maximize the margin between the two classes and to separate the two classes well. The parameter C controls the trade-off. When C is small, the margin between the two classes is large, but it may make more mistakes in training patterns. Or alternatively, when C is large, the SVM is likely to make fewer mistakes in training patterns; however, the small margin makes the network vulnerable for overfitting. Figure 5 depicts the functionality of the parameter C which has relatively large impact on the performance of the SVM. Usually, it is determined experimentally for a given problem.

To solve this optimization problem, we follow the similar procedure as before. Through constructing the Lagrange function and differentiating it, we obtain a dual problem as below: Find the Lagrange multipliers $\{\alpha_i, \ i = 1, 2, \ldots, l\}$ that minimize the objective function:

$$Q(\alpha) = \sum_{i=1}^{l}\alpha_i - \frac{1}{2}\sum_{i=1}^{l}\sum_{j=1}^{l}\alpha_i\alpha_j y_i y_j K(\mathbf{x_i}, \mathbf{x_j}). \tag{38}$$

subject to the constraints:

$$C \geq \alpha_i \geq 0\,, \tag{39}$$

$$\sum_{i=1}^{l} \alpha_i y_i = 0\,, \tag{40}$$

where $K(\mathbf{x_i}, \mathbf{x_j})$ is the kernel function:

$$K(\mathbf{x_i}, \mathbf{x_j}) \equiv \phi(\mathbf{x_i})^T \cdot \phi(\mathbf{x_j})\,. \tag{41}$$

The dual problem will become to:

$$\text{maximize: } Q(\alpha) = \sum_{i=1}^{l} \alpha_i - \frac{1}{2}\sum_{i=1}^{l}\sum_{j=1}^{l} \alpha_i \alpha_j y_i y_j K(\mathbf{x_i}, \mathbf{x_j})\,, \tag{42}$$

subject to:

$$\sum_{i=1}^{l} \alpha_i y_i = 0\,, \tag{43}$$

$$C \geq \alpha_i \alpha 0\,. \tag{44}$$

From this dual problem, the optimal weight vector, i.e., $\mathbf{w}_o$ and the optimal Lagrange multipliers, i.e., $\alpha_{o,i}$ of the optimal hyper-plane can be obtained:

$$\mathbf{w}_o = \sum_{i=1}^{l} \alpha_{o,i} y_i \phi(\mathbf{x_i})\,, \tag{45}$$

$$\sum_{i=1}^{l} \alpha_{o,i} y_i = 0\,. \tag{46}$$

The optimal hyper-plane that discriminates different classes is:

$$f(x) = \sum_{i=1}^{l} y_i \alpha_i K(\mathbf{x}, \mathbf{x_i}) + b\,. \tag{47}$$

One of the most commonly used kernel functions $\phi(\mathbf{x})$ is the polynomial kernel:

$$K(\mathbf{x}, \mathbf{x_i}) = (\mathbf{x}^T \mathbf{x_i} + 1)^p\,, \tag{48}$$

where p is a constant specified by users. Another kind of widely used kernel function is the radial basis function:

$$K(\mathbf{x}, \mathbf{x_i}) = \exp(-\gamma ||\mathbf{x} - \mathbf{x_i}||^2)\,, \tag{49}$$

where γ is also a constant specified by users. In practical application, C and the factors in the kernel function, for example, γ of the RBF kernel, are determined experimentally by users. According to its mathematic description, the structure of an SVM is shown in Fig. 6.

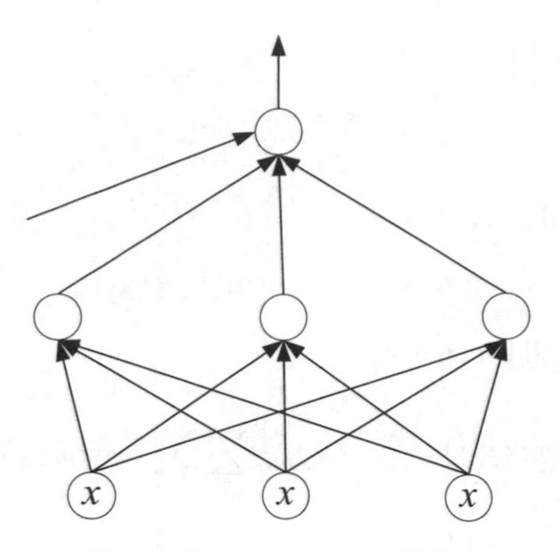

Fig. 6. The structure of an SVM.

4.3 Multi SVM classifiers

It is very common that there are more than two classes in the dataset. Therefore, binary SVMs are usually not enough to solve the whole problem. To solve multi-class classification problems, we should divide a whole problem into a number of binary classification problems. Usually, there are two approaches (Knerr et al. 1990). One is the "one against all" scheme and the other is the "one against one" scheme.

In the "one against all" scheme, if there are N classes in the entire data set, then N independent binary classifiers are built. Each binary classifier is in charge of picking out one specific class from all the other classes. For one specific pattern, all the N classifiers are used to make a prediction. The pattern is categorized to the class that receives the strongest prediction. The prediction strength is measured by the result of the decision function (47).

In the "one against one" scheme, there must be one (and only one) classifier taking charge of the classification between any two classes. Therefore, for a dataset with K classes, $K(K-1)/2$ binary classifiers are used. To get the ultimate result, a voting scheme is used. For every input vector, all the classifiers give their votes so there will be $K(K-1)/2$ votes, when all the classification (voting) finished, the vector is designated to the class getting the highest number of votes. If one vector gets highest votes for more than one class, it is randomly designated to one of them.

In fact, there is still no conclusion about which scheme is better for combining binary classifiers (Hastie and Tibshirani 1998). In our practice, we choose the "one against one" scheme.

5 Experimental Results

We applied the SVM described in the previous sections to process the SRBCT, the lymphoma, and the leukemia data sets.

5.1 Results for the SRBCT Data Set

In this data set, we first ranked the importance of all the genes with TSs. We picked out 60 of them with the largest TSs to do classification. The top 30 genes are listed in Table 2. We input these genes one by one to the SVM classifier according to their ranks. That is, we first input the gene ranked No.1 in Table 2. Then, we trained the SVM classifier with the training data and tested the SVM classifier with the testing data. After that, we repeated the whole process with top 2 genes, and then top 3

Table 2. The 30 top genes selected by t-test in the SRBCT data set.

Rank	Gene ID	Gene Description
1	810057	cold shock domain protein A
2	784224	fibroblast growth factor receptor 4
3	296448	insulin-like growth factor 2 (somatomedin A)
4	770394	Fc fragment of IgG, receptor, transporter, alpha
5	207274	Human DNA for insulin-like growth factor II (IGF-2); exon 7 and additional ORF
6	244618	ESTs
7	234468	ESTs
8	325182	cadherin 2, N-cadherin (neuronal)
9	212542	Homo sapiens mRNA; cDNA DKFZp586J2118 (from clone DKFZp586J2118)
10	377461	caveolin 1, caveolae protein, 22kD
11	41591	meningioma (disrupted in balanced translocation) 1
12	898073	transmembrane protein
13	796258	sarcoglycan, alpha (50kD dystrophin-associated glycoprotein)
14	204545	ESTs
15	563673	antiquitin 1
16	44563	growth associated protein 43
17	866702	protein tyrosine phosphatase, non-receptor type 13′ (APO-1/CD95 (Fas)-associated phosphatase)
18	21652	catenin (cadherin-associated protein), alpha 1 (102kD)
19	814260	follicular lymphoma variant translocation 1
20	298062	troponin T2, cardiac
21	629896	microtubule-associated protein 1B
22	43733	glycogenin 2
23	504791	glutathione S-transferase A4
24	365826	growth arrest-specific 1
25	1409509	troponin T1, skeletal, slow
26	1456900	Nil
27	1435003	tumor necrosis factor, alpha-induced protein 6
28	308231	Homo sapiens incomplete cDNA for a mutated allele of a myosin class I, myh-1c
29	241412	E74-like factor 1 (ets domain transcription factor)
30	1435862	antigen identified by monoclonal antibodies 12E7, F21 and O13

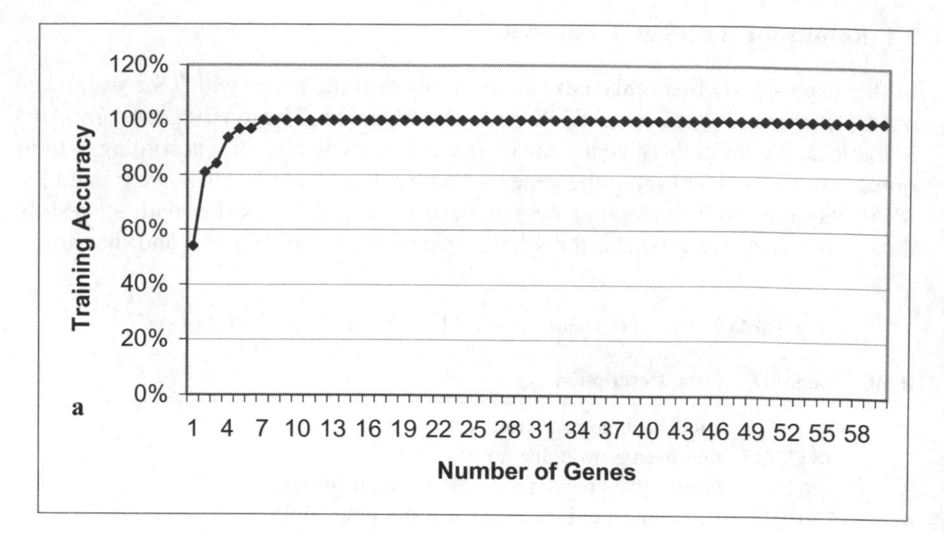

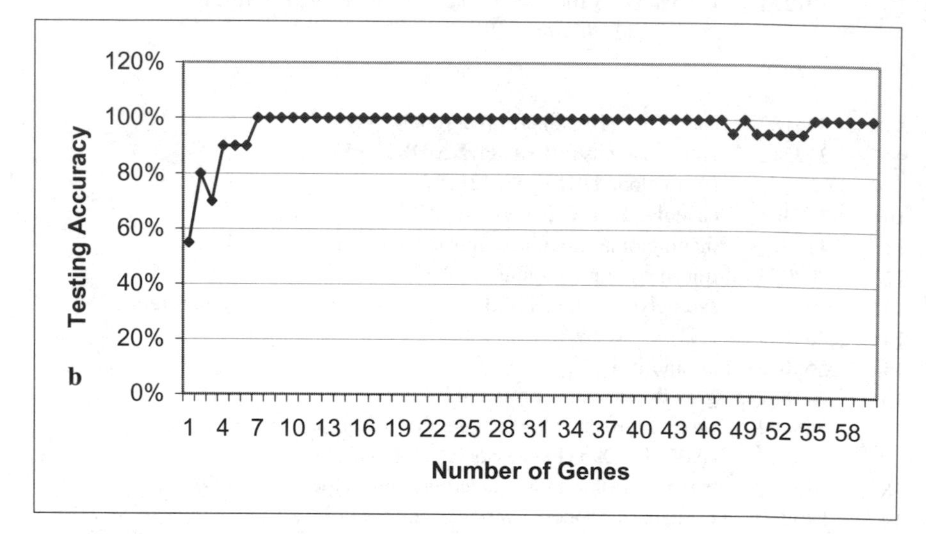

Fig. 7. The classification results for the SRBCT data set: (a) the training accuracy; (b) the testing accuracy.

genes, and so on. Figure 7 shows the training and the testing accuracies with respect to the number of genes used.

In this data set, we used SVMs with RBF kernels. C and γ were set as 80 and 0.005, respectively. This classifier obtained 100% training accuracy and 100% testing accuracy using the top 7 genes. Actually, the values of C and γ have great impact on the classification accuracy. Figure 8 shows the classification results with different values of γ (gamma). We also applied linear SVMs and SVMs with polynomial kernel function to the SRBCT data set. The results are shown in Fig. 9 and Fig. 10.

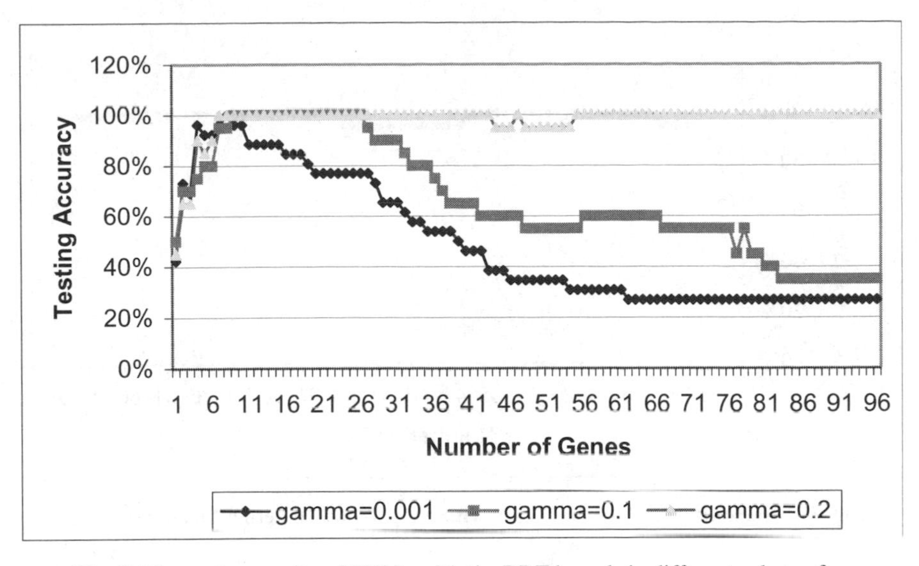

Fig. 8. The testing results of SVMs with the RBF kernels in different values of γ.

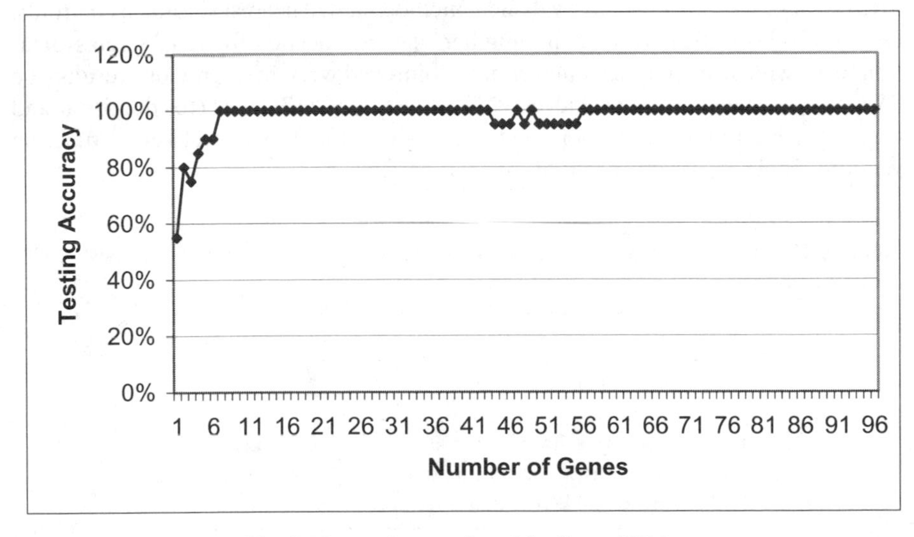

Fig. 9. The testing results of the linear SVMs.

The linear SVMs and the SVMs with the polynomial kernel function obtained 100% accuracy with 7 and 6 genes, respectively. The similarity of these results indicates that the SRBCT data set is separable for all the three kinds of classifiers.

For the SRBCT data set, Khan et al. (2001) 100% accurately classified the 4 types of cancers with a linear artificial neural network by using 96 genes. Their results and our results of the linear SVMs both proved that the classes in the SRBCT data set

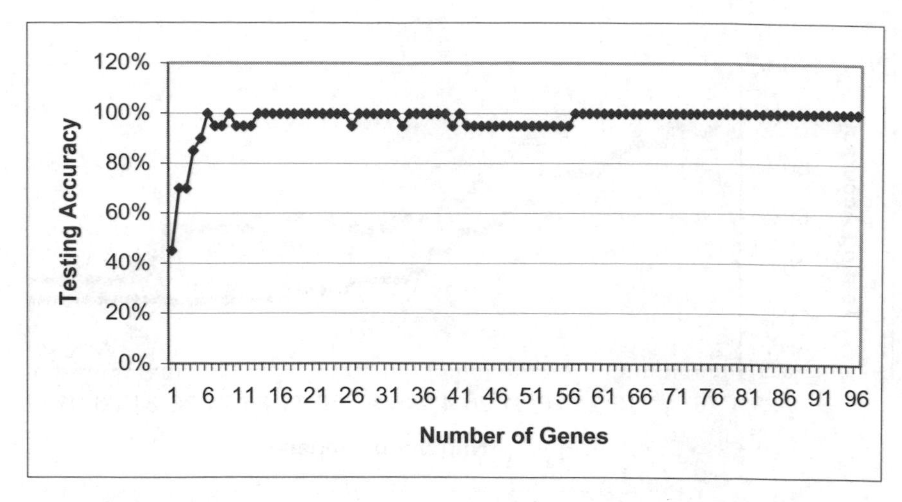

Fig. 10. The testing result of the SVMs with the polynomial kernel function ($p = 2$).

are linearly separable. In 2002, Tibshirani et al. (2002) also correctly classified the SRBCT data set with 43 genes by using a method named nearest shrunken centroids. Deutsh (2003) further reduced the number of genes needed for reliable classification to 12 with an evolutionary algorithm. Compared with these previous results, the SVMs introduced here can achieve 100% accuracy with 7 genes (for the linear and the RBF kernel function version) and 6 genes (for the polynomial kernel function version). Table 3 gives the detail of this comparison.

Table 3. Comparison of number of genes required by different methods to achieve 100% classification accuracy.

Method	Number of Genes
MLP ANN (Khan)	96
Nearest Shrunken	43
Evolutionary Algorithm	12
SVM (linear or RBF kernel function)	7
SVM (polynomial kernel function, p=2)	6

5.2 Results for the Lymphoma Data Set

In the lymphoma data set, the top 196 genes selected by TSs are listed in Table 4. Figure 12a and Figure 12b are training and testing results of the top 70 genes. The classifiers used here are also RBF SVMs. C and γ are equal to 20 and 0.1, respectively. They obtained 100% accuracy in both training and testing data set with only 5 genes.

Table 4. The top 196 important genes selected by t-test for the lymphoma data set.

Rank	Gene ID	Gene Description
1	GENE1610X	Mig=Humig=chemokine targeting T cells
2	GENE708X	Ki67 (long type)
3	GENE1622X	CD63 antigen (melanoma 1 antigen)
4	GENE1641X	Fibronectin 1
5	GENE3320X	Similar to HuEMAP=homolog of echinoderm microtubule associated protein EMAP
6	GENE707X	Topoisomerase II alpha (170kD)
7	GENE653X	Lactate dehydrogenase A
8	GENE1636X	Fibronectin 1
9	GENE2391X	Unknown
10	GENE2403X	Unknown
11	GENE1644X	cathepsin L
12	GENE3275X	Unknown UG Hs.192270 ESTs
13	GENE642X	nm23-H1=NDP kinase A=Nucleoside dephophate kinase A
14	GENE706X	CDC2=Cell division control protein 2 homolog=P34 protein kinase
15	GENE1643X	cathepsin L
16	GENE2395X	Unknown UG Hs.59368 ESTs
17	GENE537X	B-actin,1099-1372
18	GENE709X	STAT induced STAT inhibitor-1=JAB=SOCS-1
19	GENE2307X	CD23A=low affinity II receptor for Fc fragment of IgE
20	GENE2389X	Unknown
...	...	...
...	...	...
187	GENE646X	nm23-H2=NDP kinase B=Nucleoside dephophate kinase B
188	GENE2180X	Unknown
189	GENE506X	putative oral tumor suppressor protein (doc-1)
190	GENE632X	ATP5A=mitochondrial ATPase coupling factor 6 subunit
191	GENE844X	ets-2=ets family transcription factor
192	GENE629X	HPRT=IMP:pyrophosphate phosphoribosyltransferase E.C. 2.4.2.8.
193	GENE2381X	Arachidonate 5-lipoxygenase=5-lipoxygenase=5-LO
194	GENE1533X	CD11C=leukocyte adhesion protein p150,95 alpha subunit=integrin alpha-X
195	GENE2187X	SF1=splicing factor
196	GENE641X	cell cycle protein p38-2G4 homolog (hG4-1)

In the lymphoma data set, nearest shrunken centroids (Tibshirani et al. 2003) used 48 genes to give a 100% accurate classification. In comparison with this, the SVMs we used greatly reduced the number of genes required.

5.3 Results in the Leukemia Data Set

The leukemia data set is a widely known data set. Alizadeh et al. (2000) built a 50-gene classifier. This classifier made 1 error in the 34 testing samples; and in addition,

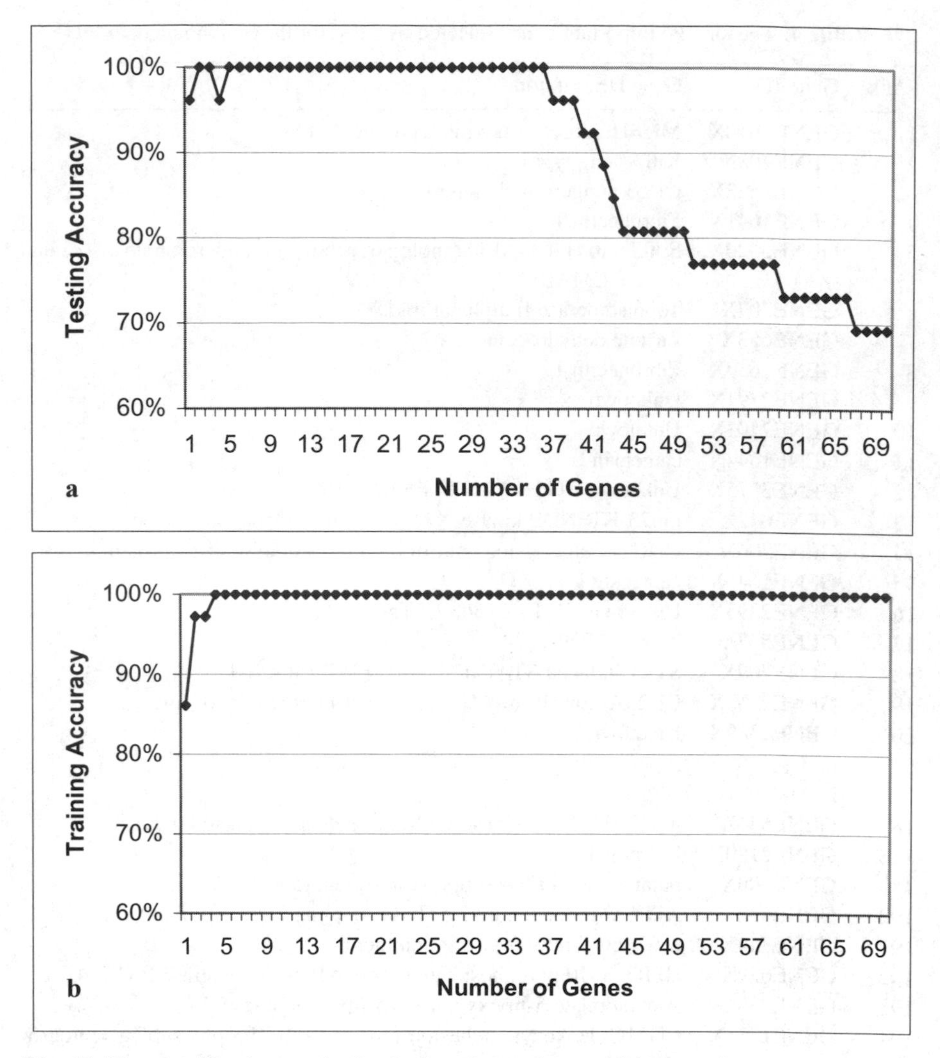

Fig. 11. The classification results for the lymphoma data set: (a) the testing accuracy; (b) the training accuracy.

it cannot give strong prediction to the other 3 samples. Nearest shrunken centroids made 2 errors among the 34 testing samples. From Fig. 12a and Fig. 12b, we know the RBF SVMs we used also made 2 errors in the testing data set. The SVMs used 20 genes to obtain this result.

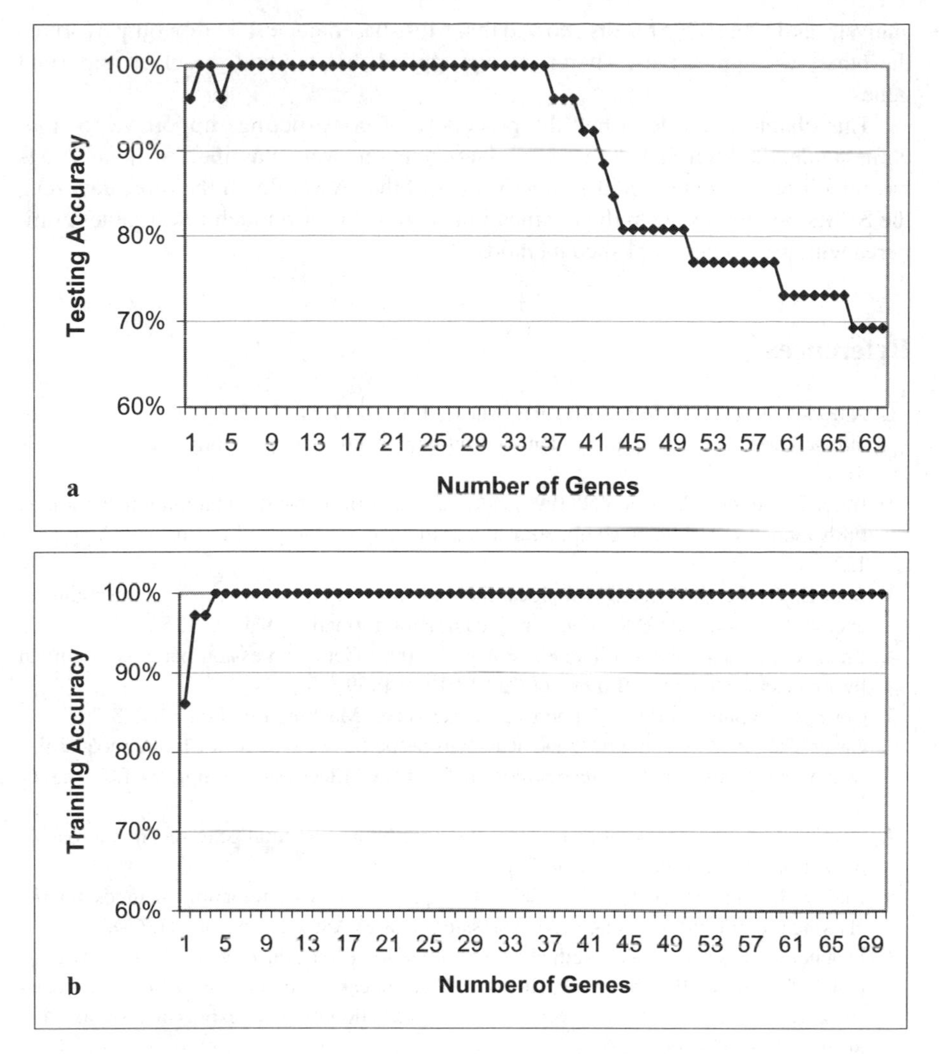

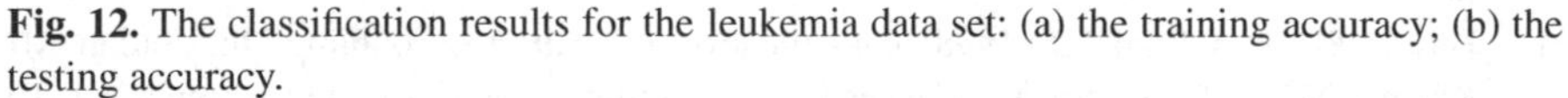

Fig. 12. The classification results for the leukemia data set: (a) the training accuracy; (b) the testing accuracy.

6 Conclusions

To find a good solution to the problem of cancer classification using gene expression data, one can work towards two related directions. One is gene selection. Selecting important genes can make the task easier because important genes determine a new input space in which the samples are more likely to be correctly classified. The other direction is to build powerful classifiers.

In this chapter, we touched upon both directions from an experimental viewpoint. For gene selection, we tested 3 well known schemes, i.e., PCA, class separability analysis, and t-test. Our results showed that t-test-based gene selection outperformed the other two approaches. Therefore, we applied this method to select important genes.

This chapter also described the procedure of constructing support vector machines in detail. After that, 3 gene expression data sets were classified with the SVMs we built. The results proved the effectiveness of the SVMs. In all the three data sets, the SVMs obtained very high classification accuracies with much fewer genes compared with previously published methods.

References

1. Alizadeh AA, Eisen MB, Davis RE, Ma C, Lossos IS, et al. (2000) Distinct types of diffuse large b-cell lymphoma identified by gene expression profiling. Nature 403:503-511
2. Boser B, Guyon I, Vapnik VN (1992) A training algorithm for optimal margin classifiers. Fifth annual workshop on computational learning theory. Morgan Kaufmann, CA pp 144-152
3. Bura E, Pfeiffer RM (2003) Graphical methods for class prediction using dimension reduction techniques on DNA microarray data. Bioinformatics 19:1252-1258
4. Chen X, Cheung ST, So S, Fan ST, Barry C (2002) Gene expression patterns in human liver cancers. Molecular Biology of Cell 13:1929-1939
5. Cortes C, Vapnik V (1995) Support vector networks. Machine Learning 20:273-297
6. Cover TM (1965) Geometrical and statistical properties of systems of linear inequalities with applications in pattern recognition. IEEE Trans. Electronic Computers EC-14:326-334
7. Deutsch J.M. (2003) Evolutionary algorithms for finding optimal gene sets in microarray prediction. Bioinformatics 19:45-52
8. Dudoit S, Fridlyand J, Speed T (2002) Comparison of discrimination methods for the classification of tumors using gene expression data. J. Am. Stat. Assoc. 97:77-87
9. Fletcher R (1987) Practical Methods of Optimization (2nd Edn.). Wiley, New York
10. Golub T, Slonim DK, Tamayo P, Huard C, Gaasenbeek M, et al. (1999) Molecular classification of cancer: class discovery and class prediction by gene expression monitoring. Science 286:531-536
11. Hastie TJ, Tibshirani RJ (1998) Classification by pairwise coupling. In: Jordan MI, Kearnsa MJ, Solla SA (eds) Advances in neural information processing systems Vol. 10, MIT Press
12. Knerr S, Personnaz L, Dreyfus G (1990) Single layer learning revisited: a stepwise procedure for building and training neural network. In: Fogelman J (eds) Neurocomputing: algorithms, architectures and applications. Springer-Verlag
13. Khan J, Wei JS, Ringner M, Saal LH, Ladanyi M, et al. (2001) Classification and diagnostic prediction of cancers using gene expression profiling and artificial neural networks. Nature Medicine 7:673-679
14. Ma XJ, Salunga R, Tuggle JT, Gaudet J, Enright E et al. (2003) Gene expression profiles of human breast cancer progression. Proc. Natl. Acad. Sci. USA. 100:5974-5979
15. Schena M, Shalon D, Davis RW, Brown PO (1995) Quantitative Monitoring of Gene Expression Patterns with a Complementary DNA Microarray. Science 267:467-470

16. Simon H (1999) Neural networks: a comprehensive foundation (2nd Edn.). Prentice-Hall, Inc. New Jersey
17. Tibshirani R, Hastie T, Narashiman B, Chu G. (2002) Diagnosis of multiple cancer types by shrunken centroids of gene expression. Proc. Natl. Acad. Sci. USA 99:6567-6572
18. Tibshirani R, Hastie,T, Narasimhan B, Chu G (2003) Class predicition by nearest shrunken centroids with applications to DNA microarrays. Statistical Science, 18:104-117
19. Troyanskaya O, Cantor M, Sherlock, G et al. (2001) Missing value estimation methods for DNA microarrays. Bioinformatics 17:520-525
20. Tusher, VG, Tibshirani R, Chu G (2001) Significance analysis of microarrays applied to the ionizing radiation response. Proc. Natl. Acad. Sci. USA 98:5116-5121
21. Vapnik VN (1998) Statistical learning theory. Wiley, New York
22. Welch BL (1947) The generalization of student's problem when several different population are involved. Biomethika 34:28-35

Artificial Neural Networks for Reducing the Dimensionality of Gene Expression Data

Ajit Narayanan, Alan Cheung, Jonas Gamalielsson, Ed Keedwell, and Christophe Vercellone

Bioinformatics Laboratory, School of Engineering, Computer Science and Mathematics, University of Exeter, Exeter EX4 4QF, United Kingdom A.Narayanan@ex.ac.uk

Summary. The use of gene chips and microarrays for measuring gene expression is becoming widespread and is producing enormous amounts of data. With increasing numbers of datasets becoming available, the need grows for well-defined, robust and interpretable methods to mine and extract knowledge from these datasets. There is currently a lot of uncertainty as to which computational and statistical methods to adopt, mainly because of the new challenges with regard to high dimensionality that gene expression data presents to the data mining community. There is a tendency for increasingly complex methods for dimensionality reduction to be proposed that are difficult to interpret. Results produced by these methods are also difficult to reproduce by other researchers. We evaluate the application of single layer, feedforward backpropagation artificial neural networks for reducing the dimensionality of both discrete and continuous gene expression data. Such networks also allow for the extraction of classification rules from the reduced data set. We demonstrate how 'supergenes' can be extracted from combined gene expression datasets using our method.

Introduction

Oligonucleotide (gene chip) and cDNA (microarray) technologies allow for the measurement of thousands of mRNAs simultaneously, providing a detailed molecular picture of cellular processes at the transcriptome level. However, interpreting the measurements is proving to be the biggest obstacle in our understanding of the transcriptome. There is considerable debate concerning the different methods used to interpret and mine the vast quantities of data produced (e.g. Brazma and Vilo, 2000; D'Haeseleer and Wen, 2000).

There are two major problems for current gene expression analysis techniques. The first and arguably the more difficult problem is that of gene dimensionality. Typically, between 7000 and 30000 genes are measured on a gene chip or microarray, and each sample or individual will therefore be associated with anything between 7000 and 30000 measurements. If there are 100 samples or individuals, the sheer volume of data leads to the need for fast analytical tools. This problem is expected to grow as gene chips develop technologically, with over 100,000 different probes being predicted for future gene chips to ensure that all splice variants of a transcriptome can be detected. The second problem is that of sparsity. While thousands of genes are measured, only a few dozen samples are taken. In database terms, there are many more attributes (genes) than records (samples). In the case of the myeloma database, for instance, there are 7129 measurements for each of only 105 individuals. There is a great danger that any technique for identifying relationships between genes and phenotype via the transcriptome will find totally spurious relationships, given the vast space of possibilities (the fewer the records, the more unconstrained the possible relationships between gene values and class).

Given these two problems, the gene expression data mining or knowledge extraction problem (**G**) can be defined to be concerned with (a) selecting a small subset of relevant genes from the original set of genes (the **S** problem) as well as (b) combining individual genes in either the original or smaller subsets of genes to identify important classificatory relationships (the **C** problem). That is, $\mathbf{G} = \mathbf{S} + \mathbf{C}$ (Figure 1).

There are numerous approaches to the **S** problem. Datasets can contain many thousands or even millions of values, hence a common strategy is simplification. Strategies for simplification include grouping and dimension reduction. Grouping

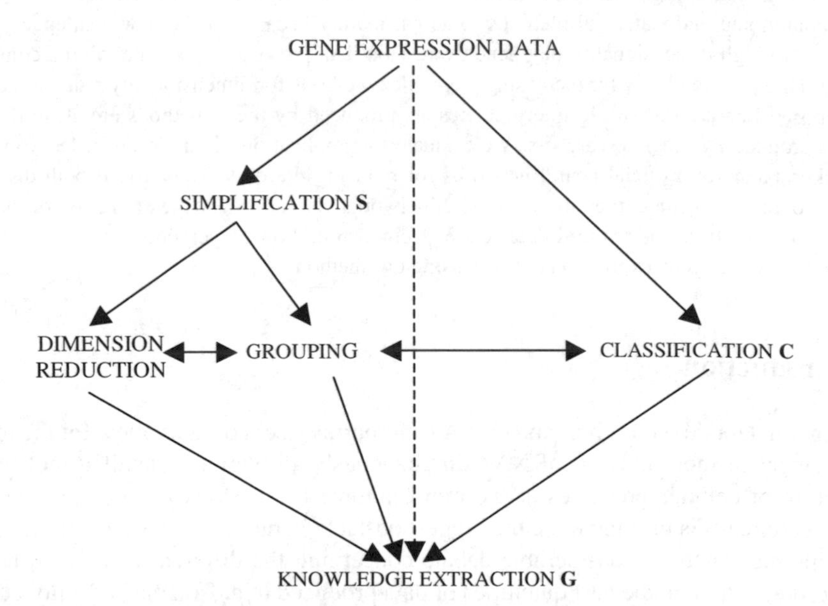

Fig. 1. Strategies for gene expression analysis. The gene expression data mining or knowledge extraction problem **G** can be defined to consist of the simplification, or gene reduction, problem **S** and/or the classification problem **C**, sometimes one following the other or at other times one being used instead of the other. It may also be possible to go from the gene expression data directly to knowledge extraction (the dotted line in the middle) without using simplification or classification. The experiments described in this paper adopt artificial neural networks for **S** followed by either the use of the most heavily weighted genes or the use of the publicly available See machine learning programme (Quinlan, 2000) for **C**.

organises the data such that similar features are shown together, reducing the complexity of the data, e.g. by clustering the genes that have similar expression profiles (e.g. Eisen *et al.*, 1998). Dimension reduction attempts to remove data that does not provide any useful information, such as genes that are not expressed or show no variation in expression between samples (e.g. Hastie *et al.* 2000). Dimension reduction can also reduce the chances of finding spurious relationships in the data. For an overview of current approaches to dimension reduction in gene expression analysis, see Baldi and Hatfield (2002).

With regard to the **C** problem, the two basic approaches are supervised and unsupervised. Supervised methods are often used to find expression patterns specific to a known phenotype or category that can then be used to accurately predict the same characteristic or category in new samples. Popular implementations of supervised learning include neighbourhood analysis (e.g. Golub *et al.*, 1999), associative networks (e.g. Bicciato *et al.*, 2001) and support vector machines (e.g. Brown *et al.*, 2000). For an overview of current supervised classificatory approaches in bioinformatics, see Narayanan et al. (2002). Unsupervised methods attempt to identify patterns within a dataset, such as sets of genes or samples sharing similar expression profiles or gene-gene interactions typical of regulatory networks of expression. No prior knowledge is assumed. Examples include hierarchical (e.g. Herrero *et al.*, 2001) and diversive clustering, self-organising maps (e.g. Tamayo *et al.*, 1999; Törönen *et al.*, 1999) and k-means clustering (Ishida *et al.*, 2002). See Baldi and Hatfield (2002), and Narayanan *et al.* (2002), for an overview of unsupervised approaches in gene expression analysis and bioinformatics generally. Novel approaches to gene expression analysis include evolutionary computation (Ando and Iba, 2001; Keedwell and Narayanan, 2003) as well as Bayesian (e.g. Baldi and Long, 2001; Friedman *et al.* 2000) and Boolean (e.g. Liang *et al.*, 1998; Shmulevich and Zhang, 2002) networks.

All these approaches have something going for them, but no method has as yet become the standard method to be adopted in the domain of gene expression analysis and modelling. One problem is that often the methods used are not easily interpretable by medical and biological experts. Complex analytical methods also present problems of repeatability and transparency. Surprisingly, there has been very little work on the application of simple supervised artificial neural networks (ANNs) for helping to solve **S** and **C**. Single layer neural networks in particular (i.e. neural networks with no hidden units) are easy to interpret, especially if one gene is associated with each input node. The influence of a gene on classification can easily be read off from the weight attached to the link between the relevant input node and the class output node.

There has been some limited application ANNs in gene expression analysis. For example, Khan *et al.* (2001) used ANNs to distinguish between four types of small round blue cell tumours. They used a strategy of dimension reduction followed by classification, using expression threshold cutoffs and principal component analysis (PCA) to reduce a 6567 gene set to 10 PCA eigenvectors. These were used as inputs into a simple perceptron of 10 input nodes and 4 output nodes. 3750 cross-validated models were produced from 63 training samples and the genes ranked according to their classification sensitivity. The top 96 genes were put through the same process again and classified 23 unseen samples with 100% accuracy. These 96 genes were also analysed using hierarchical clustering and found to cluster according to disease type. However, it is not clear why such a complex combination of PCA and ANNs is required. Also, PCA was used for dimension reduction, not ANNs.

Gruvberger *et al.* (2001) used a similar method to classify estrogen receptor (ER) positive and ER negative classes of breast cancer tumours. The method was largely the same except a multi-layer perceptron was used with 10 inputs, 5 hidden and one output node. Again, the method made near-perfect classifications. All the genes were tested for sensitivity to classification and the top 100 genes used to produce new models. Of the 58 samples, 47 were used in training and validation and 11 in testing with only 1 training sample being misclassified. Models were also produced for the top 50–150, 100–200, 150–250, 200–300, 250–350 and 300–400 genes. Although the certainty of classification was impaired, only a minor reduction in accuracy was observed. The authors suggest this is due to a highly complex gene expression pattern that is not totally governed by the top discriminator genes. However, it is more likely that this represents redundancy within the specific dataset.

O'Neill and Song (2003) use ANNs for diagnosis and prognosis of diffuse large B-cell lymphomas (DLBCL). 4026 genes from 40 patients suffering from DLBCL were classified for clinical outcome and 96 patients who donated healthy, DLBCL and non-DLBCL lymphocytes were surveyed for diagnostic classification. ANNs are used throughout their approach, both for dimension reduction and classification. Large neural networks consisting of thousands of input nodes, representing each gene, plus 100 hidden nodes were constructed and trained. Sensitivity of classification to perturbation was tested for each gene and the top ranking genes used as reduced gene sets for subsequent models. 100% accuracy was recorded when predicting the clinical outcome of 20 independent samples using a reduced set of 34 genes, and 98% accuracy in diagnosing 50 independent samples with 19 genes. Interestingly, for the diagnostic classifier, the 4026 genes were reduced to 292 and then arbitrarily partitioned (using odd and even labels) to two 146 gene sets. The even set was then reduced to 7 genes and the original 292 gene set reduced to 12 genes. These 19 genes were then used in the final models. This rather arbitrary step in the selection of genes reduces the possibility that the final gene set represents key non-redundant genes in the etiology of this cancer, despite the high accuracy. Rather, it could represent a degree of information redundancy, similar to the results obtained by Gruvberger *et al.* (2001).

The aim of the research described here is to evaluate whether a 'return-to-basics' approach with regard to ANNs will work as effectively as the more complex approaches above, all of which are difficult to reproduce independently by other researchers. Our approach is to adopt single layer ANNs, whereby the input nodes (genes) are connected directly to one output node (class) with no intervening layers, thereby making it possible to simply and effectively inspect the influence of each gene on classification (Figure 2). The data is used to train feedforward, backpropagation artificial neural networks (FFBP ANNs are considered to be the simplest type of ANN possible for supervised learning, i.e. when class information is avail-

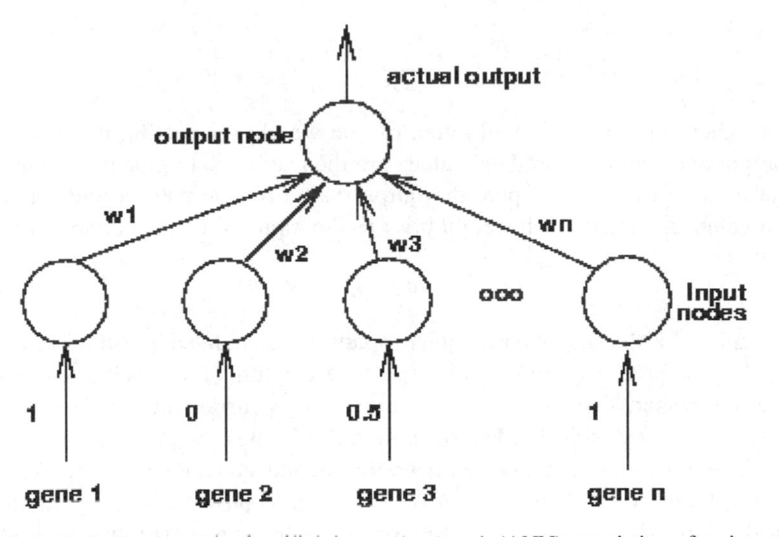

Fig. 2. A single layer, supervised artificial neural network (ANN), consisting of an input layer and an output layer. The input layer contains as many nodes as there are genes in the gene expression dataset. Absolute call values of 1 (Present), 0 (Absent) and 0.5 (Marginal) for an individual are presented to the network at the input layer (four arbitrary values are provided above). These input values are multiplied by the weights (w1 to wn) linking each input node to the output node and a weighted sum is produced at the output node ('feedforward'). This 'actual output' is compared with the desired, or 'target', output class value (1 or 0) for that individual and an error (difference between the actual output and target output) calculated. The weights are then adjusted fractionally ('learning rate') through a process known as 'back-propagation' so that the next time the same sample is input, the actual output is closer to the target value. This process is repeated for each sample in the dataset and results in one 'epoch'. The samples are input to the ANN for as many epochs as are necessary to ensure that the ANN classifies each sample accurately (typically several hundred or a few thousand epochs in our experiments).

able). Thresholds are set to provide a selection criterion for a reduced number of genes, and these genes are then used in the second iteration of the procedure. The process of training and selection is repeated to produce increasingly smaller sets of genes until a small number (typically between 20 to 40) of genes are arrived at. This handful of genes can be used to predict and classify new cases using the most strongly weighted genes or the symbolic machine learning tool, See 5 (Quinlan, 2000).

Single layer neural networks

More formally and very generally, the training phase of an ANN starts by allocating random weights $w_1, w_2, \ldots w_n$ to the connections between the n input units and the output units. Second, we feed in the first pattern p of values $x_1(p), x_2(p) \ldots x_n(p)$ to the network and compute an activation value for the output unit given p:

$$O(p) = \sum_{i=1}^{n} x_i(p)w_i(p)$$

That is, each input value is multiplied by the weight connecting its input node to the output node, and all weighted values are then summed to give us a value for the output node. Third, we compare the output value for the pattern with the desired output value and update each weight prior to the input of the next pattern p':

$$w_i(p') = w_i(p) + \Delta w_i(p)$$

where $\Delta w_i(p)$ is the weight correction for pattern p calculated as follows: $\Delta w_i(p) = x_i(p) \times e(p)$, where $e(p) = O_D(p) - O(p)$, where in turn $O_D(p)$ is the desired output for the pattern and $O(p)$ is the actual output. This is carried out for every pattern in the dataset (usually with shuffled, or random, ordering). At that point we have one epoch. The process is then repeated from the second step above for a second epoch, and a third, and so on. Typically, an ANN is said to have 'converged' or 'learned' when the sum of squared errors (SSE) on the output nodes for all patterns in one epoch is sufficiently small (typically 0.001 or below). The equations above constitute the 'delta learning rule' that can be used to train single-layer networks.

Myeloma gene expression data

Dataset

The first dataset we used contained gene expression profiles of multiple myeloma patients that were established[1] with Affymetrix Gene Chip.txt files[2] from 74 diagnosed multiple myeloma patients and 31 normal bone marrow individuals (105 samples in total) for 7129 genes. Unpublished work[3] had already identified a subset of 70 genes (1%) of the 7129 as being important for myeloma classification, based on a mixture of statistical techniques including clustering and information gain scores. However, many of these 70 genes are clearly not relevant to the disease, and some of the most important genes necessary for classification as determined by their use of See 5 appear at the very end of the list of 70 genes surrounded by irrelevant genes. If a cut-off of, say, 40 had been used, some of the most important genes would not have been included in the final list.

The data available is in the form of one file per each person. The Affymetrix process, in addition to providing real and integer numerical values for genes (AD values), also marked each gene in 'absolute call' (AC) terms: Present, Absent and

[1] The myeloma dataset is publicly available at http://lambertlab.uams.edu/publicdata.htm.

[2] More details on the Affymetrix process can be obtained from www.affymetrix.com.

[3] Page D., Zhan F., Cussens J., Waddell M., Hardin J., Barlogie B., Shaughnessy J. Jnr (2002) "Comparative data mining for microarrays: A case study based on multiple myeloma". Poster presentation at ISMB02, Edmonton. Presentation slides available at http://www.biostat.wisc.edu/~mwaddell/pres/ismb02.pdf.

Marginal. These absolute call values are produced by Affymetrix using the one-sided Wilcoxon test. Present became 1, Absent became 0, and Marginal 0.5 in our experiments.

The data were input to standard FFBP ANNs created through the SNNS software package.[4] Our hypothesis, given absolute call gene values of 1, 0 and 0.5, and class values 1 and 0, is that a negative weight value for class value 1 signifies *absence* of the gene as necessary for that class value, whereas a positive weight value for class value 1 signifies *presence* of the gene as necessary for that class value. Similarly, a negative weight value for class value 0 signifies the *presence* of that gene as necessary for that class value, and a positive weight value for class value 0 signifies *absence* of that gene as necessary for that class value. After successful training, the weights can all be compared and arbitrary thresholds set for positive and negative weights to help reduce the original gene set to a smaller gene set (i.e. genes with positive and negative weight values that exceed the thresholds are retained and the rest discarded), and the learning process repeated for all samples but using the reduced set of genes only. One implication is that successful reduction will be signified by increasingly larger weight values and fewer 'neutral' weights (weights around zero) as the process converges on an optimal set of reduced genes. All these predictions were tested in our experiments.

Experimental results

The 7129 gene expression AC values for all 105 samples were presented to an initially randomised one-layer (7129 input nodes, one output node) FFBP ANN, in shuffled order and with a learning rate of 0.001, until the sum of squared error (SSE) on the output node was below 0.001 (3000 epochs). The weights were then output and the process repeated a further two times to ensure that the random initialisation of the network led to identical or near identical weight values, which they did. The minimum weight value was −0.08196 and the maximum 0.07343, with the average being 0.000746 and the standard deviation 0.011096. It was noted that 1443 of the links consistently had 0 on their weights across all runs, signifying absolutely no effect or equal effect of 1443 genes on the two classification values 'myeloma' and 'normal'.

Arbitrary weight thresholds of −0.03 and +0.03 were used to select the top 220 genes. These gene values for each sample were extracted from the full dataset, together with the class of each sample, and the process of training the ANN was repeated for the reduced gene set, again three times to check for stability of final weights with randomised initial weights, until the SSE was below 0.001 (1000 epochs). This resulted in a network with minimum weight value of −0.2124, maximum of 0.22868, average of 0.01557 and standard deviation of 0.07295. It was noted that all genes contributed to classification, with no zero weights. Also, negatively weighted genes became even more negatively weighted, and similarly with

[4] Stuttgart Neural Network Simulator (SNNS) software is available free from http://www-ra.informatik.uni-tuebingen.de/SNNS.

Table 1. The set of weights (ordered) for the final set of 21 myeloma genes with their accession numbers is provided in the top half. Note that the left hand column contains genes with negative values (signifying that the *absence* of these genes contributes to myeloma) and that the right hand column contains genes with positive values (signifying that the *presence/over-expression* of these genes contributes to myeloma). The smaller the weight in the left hand column, the greater its contributory absence (towards the top of the column), and the greater the weight in the right hand column, the greater its contributory presence (towards the bottom of the column). The bottom half of the table provides the gene names referenced by these accession codes. See the text for an explanation of the genes in boldface and/or underlined.

U24685: −1.84127	M55267: 0.43106
L00022: −1.79993	HG491-HT491: 0.84699
Z74616: −1.63938	U09579: 0.86182
Z00010: −1.47947	HG4716-HT5158: 0.91151
U64998: −1.33768	S71043: 0.94262
K02882: −1.18341	HG2383-HT4824: 1.14876
L36033: −1.08318	L05779: 1.16866
D87024: −0.82621	Y09022: 1.19088
	U78525: 1.27232
	X57129: 1.44481
	M34516: 1.46992
	X57809: 1.58233

U24685: Human anti-B cell autoantibody IgM heavy chain variable V-D-J region (VH4) gene, clone E11, VH4-63 non-productive rearrangement;
L00022: Human Ig active heavy chain epsilon-1 gene, constant region;
Z74616: H.sapiens mRNA for prepro-alpha2(I) collagen;
Z00010: H.sapiens germ line pseudogene for immunoglobulin kappa light chain leader peptide and variable region (subgroup V kappa I);
U64998: Human ribonuclease k6 precursor gene, complete cds;
K02882: Human germline IgD-chain gene, C-region, second domain of membrane terminus;
L36033: Human pre-B cell stimulating factor homologue (SDF1b) mRNA, complete cds;
D87024: Homo sapiens immunoglobulin lambda gene locus DNA, clone:92H4;
M55267: Human EV12 protein gene, exon 1;
HG491-HT491: CD32, Fc fragment of IgG, low affinity IIa, receptor for (CD32)
U09579: Human melanoma differentiation associated (mda-6) mRNA, complete cds
HG4716-HT5158: NP_003866, Guanosine 5'-Monophosphate Synthase
L05779: Human cytosolic epoxide hydrolase mRNA, complete cds;
Y09022: H.sapiens mRNA for Not56-like protein;
U78525: Homo sapiens eukaryotic translation initiation factor (eIF3) mRNA, complete cds
X57129: H.sapiens H1.2 gene for histone H1;
M34516: Human omega light chain protein 14.1 (Ig lambda chain related) gene, exon 3
X57809: Human rearranged immunoglobulin lambda light chain mRNA

positively weighted genes. Arbitrary thresholds of −0.15 and +0.15 were set, which reduced the gene set to 21 genes.

The relevant data was extracted from the full dataset, together with the class information of each sample, and the process repeated one more time, resulting in a network that had a minimum link value of −1.84127 and a maximum of 1.58233. This final ordered set of weights for the 21 genes is contained in Table 1, together with their gene accession codes and names. It should be noted that each reduction led to a significant increase in weight value (almost ten-fold).

Lymphoma gene expression data

Dataset

The second data set consisted of frozen diagnostic nodal tumor specimens from 58 diffuse large B-cell lymphoma (DLBCL) patients that were collected and sampled.[5] DLBCL study patients who after almost five years were free of the disease (29 patients plus three additional patients who died of other causes) resulted in 32 'cured' patients, together with 23 patients who died of lymphoma plus 3 additional patients who remained alive with recurrent refractory or progressive disease (26 'fatal/refractory' patients). 7129 genes were measured using Affymetrix gene chips, and again the absolute call values were used.

Experimental results

A different procedure was adopted for the lymphoma dataset to check on the number of iterations required. A FFBP ANN was trained for all 7129 measurements for the 58 patients until the SSE on the output node was below 0.001 (5000 epochs). This resulted in an ANN with minimum weight value −0.12362 and maximum weight value of 0.12343, with average of 0.000779 and standard deviation of 0.022751. There were a high number of genes (1768) making no or equal contribution to classification (i.e. with value 0).

Given that the minimum and maximum values were much greater after the first run than for the myeloma data, it was decided to apply much higher thresholds (−0.08, +0.08) for the first cut than for the myeloma data. This resulted in a 41 gene reduced gene set. The relevant data was extracted from the full lymphoma data and re-presented to a FFBP, until the SSE on the output node was below 0.001 (1000 epochs). This resulted in a minimum weight value of −2.56309 and a maximum of 2.62468. The final ordered set of weights for the 41 genes is contained in Table 2, together with their gene accession codes.

Just as with the myeloma data, all SNNS runs were repeated to ensure that the final weight values were almost identical, irrespective of random initialisation. Again, it should be noted that the reduction led to a significant increase in weight values, except for one gene, D56495, the weight of which tended back to zero (Table 2). This may indicate that it may be safe to have two or more reductions to ensure that genes which only just made the threshold (D56495's value was 0.08553 on the first round, which is just above the threshold of 0.08) continue to have classificatory effect in subsequent rounds. Also, this confirms that not all genes will have increasing weight values just because they are in a reduced dataset.

[5] The DLBCL dataset is publicly available at http:// www-genome.wi.mit.edu/MPR/lymphoma.

Table 2. The final set of weights for the 41 genes reduced from 7219 from the lymphoma data set, ordered in terms of negative (left hand column) and positive (right hand column) weight contributions to classification (1 = fatal; 0 = cured). See the text for an explanation of D56495 (last gene in left hand column). M27830 (−2.53609) is Human 28S ribosomal RNA gene involved with transcription factors, U16306 (−2.56133) is Human chondroitin sulfate proteoglycan versican V0 splice-variant precursor peptide mRNA with some evidence in PubMed of links with brain cancers, and J02783 (2.62468) is Human thyroid hormone binding protein (p55) mRNA with some PubMed evidence of links with breast cancer. Also, see Figure 2 for gene names of genes used by See 5 for classifying lymphoma samples.

M27830	−2.56309	L37792	0.25402
U16306	−2.56133	J05272	0.74032
U24266	−2.46558	S80335	0.96776
X79781	−2.16664	HG270-HT270	1.07293
L15309	−1.88747	D87673	1.11044
U33147	−1.76716	X74614	1.11614
U66561	−1.66965	L19593	1.40816
M97676	−1.6251	S76617	1.41902
U20758	−1.50861	U16282	1.52308
HG3162-HT3339	−1.46281	HG4747-HT5195	1.71404
X65962	−1.32395	S79639	1.75022
D76444	−1.30218	U39905	1.80274
M10098	−1.21129	AB003103	1.81712
M24736	−1.02767	L15409	1.9492
D87845	−1.00562	L00058	2.08012
Y10275	−0.99936	D89289	2.10272
U96781	−0.89208	M62783	2.14169
X91196	−0.87532	U83908	2.36133
HG2197-HT2267	−0.4981	HG2090-HT2152	2.40207
U96136	−0.36178	J02783	2.62468
D56495	−0.01243		

Biological plausibility of reduced gene sets for myeloma and lymphoma

To test whether the final set of reduced genes for myeloma and lymphoma were biologically relevant, a check was made using PubMed and MedLine to identify previous references to genes in the reduced data set. For myeloma, the genes which have the largest minus and plus values are nearly all related to immunoglobulins, autoantibodies and translation initiation factors (Table 1). All 21 genes and all samples were then input to the machine learning tool, See5, to see whether See5 had enough information to distinguish myeloma patients from normal patients on the basis of these 21 genes alone. The following rule-set was produced by See5:

If L00022 (Ig active heavy chain epsilon-1) is Absent then
  If U78525 (translation initiation factor eIF3) is Absent then normal (3/1)
  Else if U78525 (translation initiation factor eIF3) is Present then myeloma (72/1)

If L00022 (Ig active heavy chain epsilon-1) is Present then
  If D87024 (Ig lambda) is Absent then myeloma (2)
  Else if D87024 (Ig lambda) is Present then normal (28)

The figures after each rule indicate the number of cases falling under this rule, and if a '/' is present, the number of false positives for that class. The overall accuracy of this rule set was a remarkable 98.1%. 74 of the 75 myeloma cases were correctly classified by these rules. A causal network representation of these rules is provided in Figure 3.

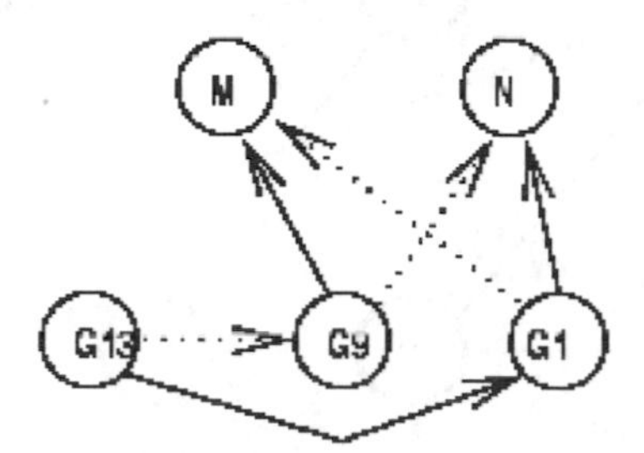

Fig. 3. Network representation of the rules extracted by See5 from the 21 lymphoma dataset genes. Unbroken lines represent 'presence' and dotted lines 'absence'. Nodes M and N represent 'myeloma' and 'normal', respectively. The interpretation of this diagram starts with the leftmost G13: 'If G13 is absent then if G9 is present then myeloma', etc. Not all absolute call values were required by See5 for classification: no 'marginal' values were involved. G13 is L00022, G9 is U78525 and G1 is D87024.

To check whether the most strongly weighted genes found by the neural network could be used for classifying the 105 cases without needing to run See5, the rules 'If U24685 (−1.84127) is absent then myeloma' and 'If L00022 (−1.79993) is absent then myeloma' (i.e. the two genes with greatest negative weight) were tested on the data. For U24685 (anti-B cell autoantibody IgM heavy chain variable V-D-J region (VH4)), 63 of the 75 myeloma cases were correctly classified, with no false positives. For L00022 (Ig active heavy chain epsilon-1), 68 of the 75 myeloma cases were covered by this rule but so were three normal cases. At the other (positive weight) extreme, X57809 (1.58233) and M34516 (1.46992) were tested. 'If X57809 (rearranged immunoglobulin lambda light chain) is present then myeloma' covered 51 myeloma cases with no false positives, and 'If M34516 (omega light chain protein 14.1 (Ig lambda chain related)) is present then myeloma' covered 61 myeloma cases with two false positives. These results provide evidence that ANN weights by themselves can be used to identify good predictors of classification. If the class value of interest is '1' (myeloma), then a negative weight is to be interpreted as the *absence* of that gene and a positive weight as the *presence*. However, all such *in silico* predictions and combinations must be tested for experimental validity as well as further biological plausibility.

For lymphoma, a more complicated set of See 5 rules emerged which is depicted as a network in Figure 4. This network classified all fatal cases (26) correctly but had one false positive for the cured class (31/1), resulting in an overall accuracy of 98.3%, again a remarkable figure. The identification of X91196 (Ataxia telangiectasia) is interesting here, since it has been previously linked with a predisposition to lymphoma (Byrd *et al.*, 1996).

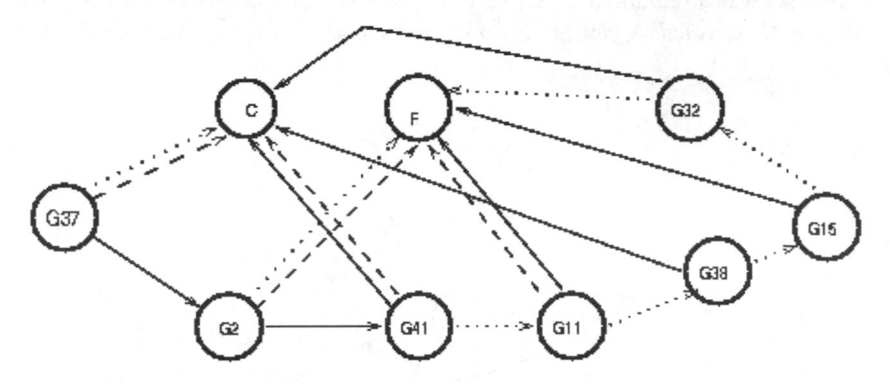

Fig. 4. Network representation of the rules extracted by See5 from the 41 lymphoma dataset genes. Unbroken lines represent 'presence', dashed lines 'marginal' and dotted lines 'absence'. Nodes C and F represent 'cured' and 'fatal', respectively. The interpretation of this diagram starts with the leftmost G37 node: 'If G37 is absent or marginal then cured'; 'If G37 is present and G2 is absent or marginal then fatal'; etc. Not all gene absolute call values were required by See5 for classification: some genes only have two arrows leaving them. G37 is HG2090-HT2152 (external membrane protein M130 antigen extracellular variant[6]). G2 is M27830 (28S ribosomal RNA gene). Gene 41 is X91196 (E14 (novel gene) and Ataxia telangiectasia proteins). G11 is HG4747-HT5195 and is a subunit of Nadh-Ubiquinone Oxidoreductase.[7] Gene 38 is HG2197-HT2267, Collage, Type Vii, Alpha 1; Gene 15 is L19593 and is Homo sapiens interleukin 8 receptor beta (IL8RB) mRNA, complete cds; and gene 32 is Y10275, H.sapiens mRNA for L-3-phosphoserine phosphatase.

Further experiments involving training and testing

The ANN experiments were repeated using a more systematic training and test method to determine the effectiveness of using ANNs for gene reduction. First, samples were divided into three equal, random, parts, with each part preserving the ratio

[6] As identified in http://www.weizmann.ac.il/physics/complex/compphys/ctwc/lgls/node13.html.

[7] As identified in http://asiago.stanford.edu/Belcher2000/table_6.html.

between the two classes. The first set is used to train an initial network. This training is made until the error is less than 0.001. Then the network is calibrated with the second set until the error is less than 0.001. The third set is used to test the entire training. This was repeated ten times. The second method involved a 'leave-one-out strategy', whereby the network is trained on all samples except one from each class, and the extracted samples are then used for testing. This was repeated five times. One other variation here was to use, as thresholds for surviving to the next round of gene reduction, one standard deviation from the extreme positive and negative values. This attempted to provide a more systematic method for selecting genes rather than the arbitrary thresholds used in the first experiments. The use of standard deviation in this manner resulted in larger numbers of genes surviving to the next round (roughly 10% each time). The results of gene reduction for only those test regimes for which the output node produced a figure of greater than 0.67 for class value 1 and below 0.33 for class value 0 were examined.

For myeloma, this resulted in a common gene reduction set (i.e. genes which appeared in every successful reduction as given by the test output) of 29 genes from the typical population of 70 to 75 genes which survived to the final stage (7129 genes were first reduced to about 700 and then to about 70 to 75 using the standard deviation method for setting thresholds). Of those 29 genes, 6 of the negatively weighted genes also appeared among the 8 negatively weighted genes from the first experiment (Table 1, in bold), and 8 of the positively weighted genes also appeared among the 12 positively weighted genes from the first experiment (Table 1, in bold). For lymphoma, unfortunately there were too few cases of successful testing using both methods to extract common genes.

Hunting for 'supergenes'

We also attempted to extract 'supergenes', i.e. genes in common to both myeloma and lymphoma. As far as we are aware, this is the first time that such an attempt has been made with ANNs. Out task was simplified by the fact that the gene chips used for myeloma and lymphoma measured the same 7129 genes. If different gene chips had been used, we would have to combine the datasets using commonly measured genes only.

First, we combined the two datasets into one file, preserving original class values (99 cases of class 1, 64 normal). After gene reduction through a FFBP ANN (first run: min −0.12342, max 0.10663, std 0.024039, threshold ±0.04 resulting in 787 genes; second run: min −0.28963, max 0.27185, std 0.1167, threshold ±0.18 resulting in 74 genes; third run: min −2.87885, max 2.79642, std 1.36488), 74 genes were extracted and run through See5, resulting in 9 rules including the following:

If M94250 and L00022 are Present then normal (27);
If M59829 is Present and X65644 and L00058 are Absent then normal (16);
If Z34897 and L00022 are Present then normal (29/1).

Overall, the accuracy of the 9 rules discovered was 96.3%. M94250 (Human retinoic acid inducible factor (MK) gene exons 1-5, with some PubMed evidence of links to carcinomas), M59829 (Human MHC class III HSP70-HOM gene (HLA), a heat shock gene), X65644 (mRNA MBP-2 for MHC binding protein 2, with some PubMed evidence that it is a transcription factor involved in the regulation of MHC class I gene expression) and Z34897 (mRNA for H1 histamine receptor, with some PubMed evidence of links to natural suppressor lymphoid cells) are genes not found in the original reductions and look promising for further evaluation.

Next, we again combined the two full datasets into one file, this time coding all lymphoma cases as 1, meaning that individuals had suffered lymphoma at some point. We adopted our refined methods above. 9 of the 10 runs of the first method (equal splitting into three data sets) and all five runs of the second method (leave one out) resulted in successful testing. When we looked at the resulting genes which were in common to all successful runs and compared them to the original myeloma and lymphoma reduced gene sets (Tables 1 and 2), interestingly only genes which appeared in Table 1 (myeloma) were supergenes for both myeloma and lymphoma. These supergenes are underlined in Table 1.

CLL and DLBCL continuous gene expression datasets

Datasets

The previous experiments involved Affymetrix absolute call (Absent, Present, Marginal) values. The next experiment was designed to test whether ANNs could successfully handle continuous gene expression data as well as discrete-valued data. The dataset was taken from the Stanford Microarray Database (SMD)[8], from previous work by Rosenwald *et al.* (2001). Lympochip cDNA microarrays (Alizadeh *et al.*, 2000; Alizadeh *et al.*, 1999) were used to generate gene expression profiles of B-cells from a variety of cancer patients including those with chronic lymphocytic leukaemia (CLL) ($n = 38$) and diffuse large B-cell lymphoma (DLBCL) ($n = 37$).

CLL is the most common human B cell leukemia and its rearranged immunoglobulin (Ig) V genes have been observed in either an unmutated or mutated state. Although both have similar clinical presentations, the former is associated with a distinctly worse prognostic outcome, and Ig V may thus be involved in two different diseases. This has been observed in DLBCL, where two prognostic classes have shown significantly different gene expression patterns (Alizadeh *et al.*, 2000; Rosenwald *et al.*, 2002). Conversely, the two CLL subtypes have been shown to have largely overlapping expression profiles and a two-group t-statistic method on $\log_2$-transformed mRNA expression ratios was used to identify the few genes that were differentially expressed (Rosenwald *et al.*, 2001).

The original dataset was partitioned into two datasets: (a) CLL and DLBCL were used to represent separate classes each, and (b) CLL was itself divided into two

[8] URL: http://genome-www.stanford.edu/microarray

classes, one representing samples with mutated Ig V genes ($n = 12$) and the other with Unmutated Ig V genes ($n = 16$). The first dataset was assumed to contain large differences between the classes, as CLL and DLBCL represent two very different types of B cell cancer, with different clinical presentations and outcomes. The second dataset was assumed to contain more subtle differences and be harder to classify.

Data was downloaded in SMD format[9] as a tab-delimited text file. All non-flagged array elements were filtered by excluding those fluorescence values lower than 1.4 times the local background, in both Cy3 and Cy5 channels (representing mRNA pooled from nine lymphoma cell lines and the CLL/DLBCL experimental samples respectively). Array elements had to pass this requirement on at least 70% of the samples, or were otherwise excluded. The normalised log2 mean expression ratios were extracted to tables representing each class. Replicates were identified with their accession numbers and a median average taken. SNNS cannot tolerate null values hence missing expression ratios were estimated using *KNNimpute* (Troyanskaya *et al.*, 2001), a commonly used tool in computing missing gene expression values.

Experimental results

The completed dataset was randomly split into a training and test set. Ten test samples (out of 75) were withheld for the DLBCL/CLL distinction and four (out of 28) for the Ig-mutated CLL/Ig-unmutated CLL distinction. This was repeated ten times so as to produce ten different train/test sets. The training data was classed as 0 for the CLL and CLL-Ig-mutated class or as 1 for the DLBCL and CLL-Ig-unmutated class before being fed into a standard feedforward back-propagation neural network created in SNNS, in shuffled order and with a learning rate of 0.0001, for a minimum of 1000 epochs until the sum of squared error was below 0.1. A classification threshold for the output neuron was set at 0.5, with values below representing class 0 and values above representing class 1. Weights were set to fall within the range -1 to 1. The trained network was evaluated with the withheld test data and checked for over-training. The connection weights between each input neuron (gene) and the output neuron were tabulated and thresholds set as the mean $\pm$ the standard deviation. Genes falling between these limits were excluded and the process repeated five times with the remaining data. The method was run for six training rounds on each dataset, resulting in five successive gene reduction steps. Results are summarised in Table 3.

Dataset 1: DLBCL vs. CLL

The first dataset has already been shown to contain clear differences in expression profiles between the two classes (Rosenwald *et al.*, 2001). As such, it was expected

[9] Database schema available from http://genome-www5.stanford.edu/cgi-bin/SMD/tableSpecifications?table=EXPERIMENT

that the neural network would perform well on this classification and it subsequently produced a near perfect classification when applied to test data. On the first three iterations, a large majority of the weights were found to have values close to zero, with a mean of -2.5×10^{-5} and a SD of 0.0024 after the first training round. The resulting threshold levels excluded 87.5% of the initial dataset, leaving 943 genes. The fact that this classification accuracy improved to 100% after the first and subsequent gene reductions is encouraging and gave confidence in this method's ability to select genes important in differentiating two distinct classes. This was backed up by cluster analysis of each reduced gene set, which showed clear patterns of differentially expressed genes, with well-defined separation of clusters representing DLBCL and CLL classes (unpublished data). A logarithmic increase in the standard deviation was observed with each gene reduction. The classification error increased after the gene set was reduced for a fifth time. By this point, the mean had increased to 0.3731 and the SD to 0.2092. The highest and lowest weights remain roughly equally sized until the fifth training round where the positive weights are considerably larger.

Table 3. The results on the CLL vs. DLBCL datasets (above) and Ig-Mutated CLL vs. Ig-Unmutated CLL data (below). Repeated applications of the ANN reveal increasing weight values as the number of genes is reduced. After each training round, test (unseen) data were evaluated against the trained ANN and the misclassifications recorded. The results indicate that best generalization on the test data is not necessarily achieved by gene greatest reduction.

Training Round	**1**	**2**	**3**	**4**	**5**	**6**
CLL (class 0) vs. DLBCL (class 1)						
Genes	7583	943	335	99	42	11
Highest Weight	0.0139	0.02060	0.03474	0.07023	0.2269	0.6865
Lowest Weight	-9.707×10^{-03}	-0.01879	-0.03086	-0.0779	-0.12888	0.08700
Mean	-2.504×10^{-05}	4.43×10^{-04}	7.814×10^{-04}	6.788×10^{-03}	-0.02615	0.3731
SD	2.405×10^{-03}	8.500×10^{-03}	0.01754	0.04754	0.1128	0.2092
Misclassifications	2/10	0/10	0/10	0/10	0/10	5/10
Ig-Mutated CLL (class 0) vs. Ig-Unmutated CLL (class 1)						
Genes	10492	2452	627	164	58	22
Highest Weight	0.02881	0.03583	0.06400	0.1265	0.2254	0.4637
Lowest Weight	-0.01836	-0.02272	-0.03800	-0.07867	-0.1342	-0.2769
Mean	-4.7189×10^{-05}	-2.706×10^{-05}	8.560×10^{-04}	-3.4×10^{-04}	0.01939	-0.06932
SD	2.603×10^{-03}	5.971×10^{-03}	0.01601	0.04559	0.1014	0.2342
Misclassifications	2/4	0/4	0/4	0/4	0/4	2/4

Dataset 2: CLL vs. CLL

This dataset contained much subtler differences so it was a surprise to find that the neural network provided as accurate a classification of test data as the previous dataset. However, the sample sizes were much smaller (as were the number of test cases) and as such, perfect classification accuracy should be viewed with some caution. As with the previous dataset, classification accuracy began to suffer after the final gene reduction. Many trends from the previous dataset are repeated here, with average weight and SD increasing with each iteration (to final values of –0.06932 and 0.2342 respectively). It was encouraging to see many of the genes identified in the original study along with previously unidentified genes, perhaps potential candidates for further study (Table 4, Figure 5). ZAP-70 (AI364932) was identified in the original study (Rosenwald *et al.*, 2001) as the most discriminatory gene between the two classes. Subsequent studies have measured ZAP-70 expression with flow cytometry, western blotting and immunohistochemistry (Crespo *et al.*, 2003; Wiestner *et al.*, 2003) and have shown it to be well correlated with Ig mutational status and probability of survival. Whilst ZAP-70 was found in this study at every stage of gene reduction (with a weight of 0.2494 after the final training round), it was not rated as the highest weighted gene. Activation Induced C-type lectin (AICL) was rated as the most discriminant gene, with a weight of 0.4637. AICL is encoded in the NK gene complex, found to be upregulated in activated lymphocytes and is presented at the cell surface (Hamann *et al.*, 1997). This could be an alternative target for further study.

Throughout the training, the highest weights were significantly larger than the lowest values, and becoming more pronounced with each gene reduction, suggest-

Table 4. Top 22 ranked genes (with weights) from Dataset 2 (Ig-Mutated CLL vs. Ig-Unmutated CLL).

Weight	Gene
-0.276846	AI434176 Unknown
-0.247223	AA459003 myb-related_gene_A_A-myb
-0.230393	AA814450 Unknown_UG_Hs.63671_ESTs
-0.215547	AA836308 Unknown
-0.211463	AI095207 Unknown
-0.208339	AA283119 APR_immediate-early-response_gene_ATL-derived_PMA-responsive
-0.195464	AA748744 Unknown_UG_Hs.181384_ESTs__sc_id5564
-0.195291	AA836095 myb-related_gene_A_A-myb
-0.191631	AA814334 Unknown
-0.187753	AA262420 APR_immediate-early-response_gene_ATL-derived_PMA-responsive
-0.184405	AA283601 Unknown
-0.167194	NO_ACC_47
-0.160495	AA747694 Unknown_UG_Hs.159556_ESTs_Weakly_similar_to_ALU_SUBF
-0.157318	AA243356 Unknown_UG_Hs.88102_ESTs
-0.138412	N30237 NKG2-D_type_II_integral_membrane_protein_on_human_natural_ki
-0.134013	AA504350 myb-related_gene_A_A-myb
-0.114734	AA731747 titin
0.220947	AA812170 BCL-7A
0.249423	AI364932 ZAP-70
0.315697	AA417616 BCL-7A
0.441694	Lipoprotein_lipase
0.463691	H11732 AICL_activation-induced_C-type_lectin

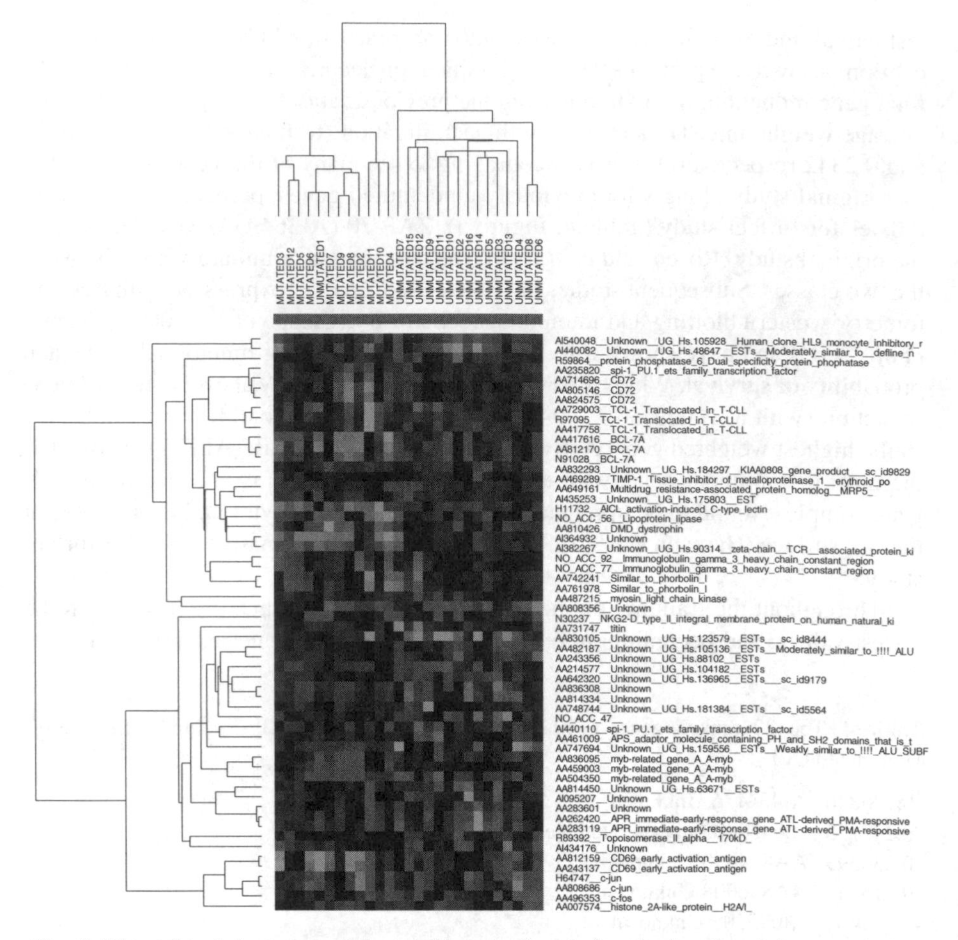

Fig. 5. Hierarchical clustering of the top 54 genes found after the fourth training round (third gene reduction) that most strongly discriminated between the two CLL subtypes. Note that one unmutated sample falls within the mutated class (top of diagram).

ing that genes upregulated in Ig-unmutated CLL contribute more to the classification than downregulated genes. The distribution of gene weights was roughly equal between positive and negative values for the first four training rounds although a bias toward positively weighted genes was seen in the fifth training round. This reverts in the sixth training round with an obvious negative bias.

Discussion and conclusion

The application of simple, feedforward backpropagation neural networks (FFBP ANNs) to the problem of reducing the dimensionality of gene expression data has resulted in smaller sets of genes with increasingly larger weight values which, upon subsequent analysis, have been demonstrated to be strongly related to these diseases in exactly the way that the weight values would predict (presence and absence). The results of See5 are almost perfect for the reduced gene sets, indicating that with some fine-tuning perfect classification can be achieved after gene reduction. It is also possible that new discoveries concerning the involvement of specific genes not previously reported in the literature will arise. Comparisons with previous data reduction techniques used for the myeloma data indicate that the use of FFBP ANNs identifies a smaller and more relevant gene set.

The application of See5 provides a powerful check on the results of the data reduction and leads directly to the generation of candidate classificatory diagrams for further analysis. However, we can also derive classificatory rules directly from the trained ANNs if we wish. Our approach can be applied equally to discrete as well as continuous gene expression data. Finally, our approach can be applied to the search for supergenes.

No detailed knowledge of statistics is required with our tool-based approach. In other words, we have provided evidence that gene expression analysis can be undertaken with simplest gene reduction techniques, in this case, one-layer neural networks that essentially perform linear discriminant or linear regression analysis. However, the use of ANNs provides powerful visual tools for gene expression analysts as well as potential for easy repeatability of experiments. And the application of See5 produces results which are immediately interpretable by domain experts. In conclusion, our experiments demonstrate that, just because the gene reduction problem is complex, this does not mean that the solution has to be complex also.

References

1. Alizadeh, A. A., Eisen, M. B., Davis, R. E., Ma, C., Lossos, I. S., Rosenwald, A., Boldrick, J. C., and Sabet, H. (2000). Distinct types of diffuse large B-cell lymphoma identified by gene expression profiling. *Nature* **403**, 503-511.
2. Alizadeh, A. A., Eisen, M. B., Davis, R. E., Ma, C., Sabet, H., Tran, T., Powell, J., Yang, L., Marti, G. E., and Moore, T. (1999). The Lymphochip: a specialised cDNA microarray for the genomic-scale analysis of gene expression in normal and malignant lymphocytes. *Cold Spring Harbor Symp. Quant. Biol.* **64**, 71-78.
3. Ando, S. and Iba, H. (2001) Inference of gene regulatory models by genetic algorithms. *Proceedings of the 2001 IEEE Congress on Evolutionary Computation*: 712-719.
4. Baldi, P. and Hatfield, G. W. (2002) *DNA Microarrays and Gene Expression*. Cambridge University Press.
5. Baldi, P. and Long, A. D. (2001) A Bayesian framework for the analysis of microarray expression data: Regularised t-test and statistical inferences of gene change. *Bioinformatics* 17: 509-519.
6. Bicciato, S., Pandin, M., Didonè, G. and Di Bello, C. (2001) Analysis of an associative memory neural network for pattern identification in gene expression data. *Workshop on Data Mining in Bioinformatics (BIOKDD01)*. Available from http://www.cs.rpi.edu/~zaki/BIOKDD01/

7. Brazma, A. and Vilo, J. (2000) Gene expression data analysis. *FEBS Letters* 480 (1): 17-24. Also available at http://industry.ebi.ac.uk/~vilo/Publications/Febs_Brazma_Vilo_00023893.ps.gz
8. Brown, M.P.S., Grundy, W.R., Lin, D., Cristianini, N., Sugnet, C., Furey, T.S., Ares, M. Jr. and Haussler, D. (2000) Knowledge-based analysis of microarray gene expression data using support vector machines. *Proceedings of the National Academy of Sciences* 97(1): 262-267.
9. Byrd, P.J., McConville, C.M., Cooper, P., Parkhill, J., Stankovic, T., McGuire, G.M., Thick, J.A. and Taylor, A.M. (1996). Mutations revealed by sequencing the 5′ half of the gene for ataxia telangiectasia. *Human Molecular Genetics* 5(1): 145-149.
10. Crespo, M., Bosch, F., Villamor, N., Bellosillo, B., Colomer, D., Rozman, M., Marcé, S., López-Guillermo, A., Campo, E., and Montserrat, E. (2003). ZAP-70 Expression as a Surrogate for Immunoglobulin-Variable-Region Mutations in Chronic Lymphocytic Leukemia. *N. Engl. J. Med.* **348**, 1764-1775.
11. D'haeseleer, P. and Wen, X. (2000) Mining the gene expression matrix: Inferring gene relationships from large scale gene expression data. *Bioinformatics* 16 (8): 707-726 (available from www.bioinformatics.oupjournals.org).
12. Eisen, M. B., Spellman, P. T., Brown, P. O., Botstein, D. (1998) Cluster analysis and display of genome-wide expression patterns. *Proceedings of the National Academy of Sciences USA* 95 (25): 14863-14868.
13. Friedman, N., Linial, M., Nachman, I. and Pe'er, D. (2000) Using Bayesian network to analyze expression data. *Journal of Computational Biology* 7: 601-620.
14. Golub, T.R., Slonim, D.K., Tamayo, P., Huard, C., Gaasenbeek, M., Mesirov, J.P., Coller, H., Loh, M.L., Downing, J.R., Caligiuri, M.A., Bloomfield, C.D. and Lander, E.S. (1999) Molecular classification of cancer: Class discovery and class prediction by gene expression monitoring. *Science* 286, 531-537.
15. Gruvberger, S., Ringner, M., Chen, Y., Panavally, S., Saal, L.H., Borg, A., Ferno, M., Peterson, C. and Meltzer, P.S. (2001) Estrogen Receptor Status in Breast Cancer Is Associated with Remarkably Distinct Gene Expression Patterns. *Cancer Research* 61: 5979-5984.
16. Hamann, J., Montgomery, K. T., Lau, S., Kucherlapati, R., and van Lier, R. A. (1997). AICL: a new activation-induced antigen encoded by the human NK gene complex. *Immunogenetics* **45**, 295-300.
17. Hastie, T., Tibshirani, R., Eisen, M. B., Alizadeh, A., Levy, R., Staudt, L., Chan, W. C., Botstein, D. and Brown, B. (2000) Gene shaving as a method for identifying distinct sets of genes with similar expression patterns. *Genome Biology* 1(2):research0003.1-0003.21.
18. Herrero, J.,Valencia, A. and Dopazo, J. (2001) A hierarchical unsupervised growing neural network for clustering gene expression patterns. *Bioinformatics* 17 (2): 126-136.
19. Ishida, N., Hayashi, K., Hoshijima, M., Ogawa, T., Koga, S., Miyatake, Y., Kumegawa, M., Kimura, T. and Takeya, T. (2002) Large scale gene expression analysis of osteoclastogenesis in vitro and elucidation of NFAT2 as a key regulator. *Journal of Biological Chemistry* 277(43):41147-56.
20. Keedwell, E. C. and Narayanan, A. (2003) Genetic algorithms for gene expression analysis. *Applications in Evolutionary Computing: Proceedings of the 1st European Workshop on Evolutionary Bioinformatics*, Raidl *et al.* (Eds.), Springer: 76-86.
21. Khan, J., Wei, J. S., Ringner, M., Saal, L. H., Ladanyi, M., Westerman, F., Berthold, F., Schwab, M., Antonescu, C. R., Peterson, C., and Meltzer, P. S. (2001) Classification

and diagnostic prediction of cancers using gene expression profiling and artificial neural networks. *Nature Medicine* 7: 673-679.

22. Liang, S., Fuhrman, S. and Somogyi, R. (1998) REVEAL, A General Reverse Engineering Algorithm for Inference of Genetic Architectures. *Pacific Symposium on Biocomputing* 3: 18-29.
23. Narayanan, A., Keedwell, E. C. and Olsson, B. (2002) Artificial intelligence techniques for bioinformatics. *Applied Bioinformatics* 1(4): 191-222.
24. O'Neill, M. C. and Song, L. (2003) Neural network analysis of lymphoma microarray data: prognosis and diagnosis near-perfect. *Bioinformatics* 4 (1): 13.
25. Quinlan, J. R. (2000) Data mining tools See5 and C5.0. http://www.rulequest.com/see5-info.html.
26. Rosenwald, A., Alizadeh, A. A., Widhopf, G., Simon, R., Davis, R. E., Yu, X., Yang, L., Pickeral, O. K., Rassenti, L. Z., Powell, J., Botstein, D., Byrd, J. C., Grever, M. R., Cheson, B. D., Chiorazzi, N., Wilson, W. H., Kipps, T. J., Brown, P. O. and Staudt, L. (2001). Relation of gene expression phenotype to immunoglobulin mutation genotype in B cell chronic lymphocytic leukemia. *J. Exp. Med.* **194**, 1639-1647.
27. Rosenwald, A., Wright, G., Chan, W. C., Connors, J. M., Campo, E., Fischer, R. I., and Gascoyne, R. D. (2002). The use of molecular profiling to predict survival after chemotherapy for Diffuse Large B-Cell Lymphoma. *N. Engl. J. Med.* **346**, 1937-1947.
28. Shmulevich, I. and Zhang, W. (2002) Binary analysis and optimisation-based normalization of gene expression data. *Bioinformatics* 18 (4): 555-565.
29. Tamayo, P., Slonim, D., Mesirov, J., Zhu, Q., Kitareewan, S., Dmitrovsky, E., Lander E. S. and Golub, T. R. (1999) Interpreting patterns of gene expression with self-organising maps: methods and applications to hematopoietic differentiation. *Proceedings of the National Academy of Sciences USA* 96: 2907-2912.
30. Törönen, P., Kolehmainen, M., Wong, G. and Castrén, E. (1999) Analysis of gene expression data using self-organising maps. *FEBS Letters* 451: 142-146.
31. Troyanskaya, O., Cantor, M., Sherlock, G., Brown, P., Hastie, T., Tibshirani, R., Botstein, D., and Altman, R. B. (2001). Missing value estimation methods for DNA microarrays. *Bioinformatics* 17, 520-525.
32. Wiestner, A., Rosenwald, A., Barry, T. S., Wright, G., Davis, R. E., Henrickson, S. E., Zhao, H., Ibbotson, R. E., Orchard, J. A., Davis, Z., Stetler-Stevenson, M., Raffeld, M., Arthur, D. C., Marti, G. E., Wilson, W. H., Hamblin, T. J., Oscier, D. G., and Staudt, L. M. (2003). ZAP-70 expression identifies a chronic lymphocytic leukemia subtype with unmutated immunoglobulin genes, inferior clinical outcome, and distinct gene expression profile. *Blood* **101**, 4944-4951.

Printing: Strauss GmbH, Mörlenbach
Binding: Schäffer, Grünstadt